# Dogs That Know When Their Owners Are Coming Home

Also by Rupert Sheldrake

*Morphic Resonance*

*A New Science of Life*

*The Presence of the Past*

*The Rebirth of Nature*

*The Sense of Being Stared At*

*Seven Experiments That Could Change the World*

with Ralph Abraham and Terence McKenna

*The Evolutionary Mind*

*Trialogues at the Edge of the West*

with Matthew Fox

*Natural Grace*

*The Physics of Angels*

# Dogs That Know When Their Owners Are Coming Home

## And Other Unexplained Powers of Animals

Fully updated and revised

Rupert Sheldrake

THREE RIVERS PRESS • NEW YORK

Published in the United States by Three Rivers Press, an imprint of the Crown Publishing Group, a division of Random House, Inc., New York.
www.crownpublishing.com

Three Rivers Press and the Tugboat design are registered trademarks of Random House, Inc.

A previous edition of this work was published in hardcover in the United States by Crown Publishers, an imprint of the Crown Publishing Group, a division of Random House, Inc., New York, in 1999, and was subsequently published in paperback in the United States by Three Rivers Press, an imprint of the Crown Publishing Group, a division of Random House, Inc., New York, in 2000.

Library of Congress Cataloging-in-Publication Data

Sheldrake, Rupert.
Dogs that know when their owners are coming home : and other unexplained powers of animals / Rupert Sheldrake.—1st rev. ed.
p. cm.
Includes bibliographical references and index.
1. Pets—Psychic aspects. 2. Extrasensory perception in animals. I. Title.
SF412.5.S5 2011
636.088'7—dc22
2011004065

ISBN 978-0-307-88596-8
eISBN 978-0-307-88846-4

Printed in the United States of America

Book design by Chris Welch
Cover design by Jennifer O'Connor
Cover photography: IPS Co., Ltd./Corbis (dog); © Laura Doss/Brand X/Corbis
Back cover photographs: iStock

1 3 5 7 9 10 8 6 4 2

First Revised Edition

*With thanks to all the animals from whom I have learned.*

# Contents

# Preface

This is a book of recognition—a recognition that animals have abilities that we have lost. One part of ourselves has forgotten this; another part has known it all along.

As a child, like many other children, I was interested in animals and plants. My family kept a great variety of pets. In addition to our dog, Scamp, we had a rabbit, hamsters, pigeons, a jackdaw, a budgerigar, a terrapin, two tortoises, several goldfish, and tadpoles and caterpillars I would rear each spring. My father, Reginald Sheldrake, a pharmacist and amateur microscopist, encouraged my interests and fueled my fascination with the natural world when he showed me how drops of pond water teemed with myriad forms of life, and what the scales on butterflies' wings looked like.

I was especially intrigued by the way that pigeons homed. On Saturday mornings my father took me to see a great liberation of them. At our local railway station at Newark-on-Trent, in the English Midlands, racing birds from all over Britain were waiting in wicker baskets, arrayed in stacks. At the appointed time, the porters let me help them open the flaps. Out burst hundreds of pigeons in a great commotion of wind and feathers. They flew up into the sky, circled around, and set off in various directions toward their faraway homes. How did they do it? I wondered. No one seemed to know. Their homing ability is still unexplained today.

At school it was a natural choice for me to study biology and other sciences, and I continued my scientific education at Cambridge University, where I studied botany, physiology, chemistry, and biochemistry as an undergraduate, and then took a Ph.D. in biochemistry. But as I proceeded in my education as a biologist, a great gulf began to open up between my own experience of animals and plants and the scientific approach that I was being taught.

The mechanistic theory of life, still the dominant orthodoxy, asserts that living organisms are nothing but complex genetically programmed machines. They are supposed to be inanimate, literally soulless. As a general rule, the first step we took when studying living organisms was to kill them or cut them up. I spent many hours of laboratory work in dissection, and then as my studies proceeded, in vivisection. For example, it was an essential part of my biology curriculum to dissect nerves from the severed legs of frogs and stimulate them electrically to make the muscles twitch. For the study of enzymes in rat liver, one of the favored tissues in animal biochemistry, we first had to decapitate the living rats, their blood spurting down the laboratory sink. I heard nothing about how pigeons homed.

A love of animals had led me to study biology, and this was where it had taken me. Something had gone wrong. I began to wonder what was going on, and tried to find out. After my undergraduate studies at Cambridge, I was awarded a Frank Knox Fellowship at Harvard, where I studied philosophy and the history of science, in search of a wider perspective. I then returned to Cambridge to begin research on plants.

For ten years I did research at Cambridge in developmental biology, while continuing to think about the outlines of a more holistic science. I became a Fellow of Clare College, Cambridge, where I was Director of Studies in biochemistry and cell biology. While working in Cambridge, I was elected a Research Fellow of the Royal Society; and under the auspices of the Royal Society I did research at the University of Malaya on rain-forest plants. I later became Principal Plant Physiologist at ICRISAT, the International Crops Research Institute for the Semi-Arid Tropics, in Hyderabad, India, helping to improve the growth and yield of crops that are a vital part of the diet of hundreds of millions of people.

I have spent more than forty years as a professional scientist, publishing papers in scientific journals and speaking in scientific congresses, and have long been a member of scientific societies, such as the Society for Experimental Biology, and a Fellow of the Zoological Society. I am a great believer in the value of scientific inquiry, but I am more convinced than ever that the mechanistic theory of nature is too narrow. I have discovered that an increasing number of my scientific colleagues agree, although most are reluctant to say so in public. I have found that the split I experienced within myself, the gulf between personal experience of life and the theory that living organisms, including ourselves, are merely soulless automata is widespread within and outside the scientific community.

I have come to realize that this split is not inevitable, and that a more inclusive kind of science is possible, as well as cheaper. But it is inevitably controversial. For some scientists, the mechanistic theory of nature is not just a testable hypothesis, but more like a religious creed. For others, open-minded inquiry is more important than the defense of long-entrenched dogmas. I have found such scientists a great help in my researches, and have received much encouragement and practical support.

In 1994 I published a book called *Seven Experiments That Could Change the World*[1] in which I explored seven well-known but little understood phenomena, and suggested how inexpensive research could lead to major breakthroughs. One of these experiments concerned the possible telepathic abilities of dogs and cats. In particular, I focused on the ability of some dogs to know when their owners are coming home.

Thus through trying to find ways in which a broader view of life can be developed scientifically, I have come back to pets. It took me a long time to recognize that they are the animals we know best. I knew this as a child. To many people it is blindingly obvious, but for me it had all the force of a new discovery. I realized that the animals we know best have much to teach us. They can help enlarge our understanding of life; they are not just cute, cuddly, comforting, and fun.

For five years before the first edition of this book was published in 1999 I researched the perceptiveness of pets, with the help of more than two thousand animal owners and trainers. I surveyed more than a thousand randomly chosen pet owners to find out how common various kinds of unexplained behavior are. My associates and I interviewed hundreds of people with much experience of animals, including dog trainers, search-and-rescue dog handlers, police dog handlers, blind people with guide dogs, veterinarians, kennel and stable proprietors, horse trainers and riders, farmers, shepherds, zookeepers, pet shop proprietors, reptile breeders, and pet owners.

If I had quoted from all of the accounts and interviews that I have been given, this book would have been at least ten times thicker. In some instances hundreds of people have told me about very similar patterns of behavior in their pets, like dogs knowing when their owners are coming home. I have had to condense this information, giving only a few examples of each kind of perceptive behavior. Although many people have contributed to the overall picture, I can acknowledge only a small minority by name. Without all this help from people named and unnamed, this book could not have been written. I am indebted to all those who have helped me, and to their animals.

Since the first edition of this book was published I have received more than 1,500 further reports about perceptive behavior by animals, and my database now contains more than 4,500 case histories. I have included some of them in this new edition. I have updated the text throughout and included summaries of recent scientific research on animal domestication, animal navigation, and other relevant subjects. I have also included the results of new experimental studies I have carried out with dogs and other animals, the most notable being a study with a language-using parrot, N'kisi, which showed that this amazing bird responded to his owner's thoughts at a distance and actually said what she was thinking. I also summarize some of my recent research on human telepathy, especially in connection with telephone calls. There is a fuller discussion of my research on unexplained human abilities in my book *The Sense of Being Stared At* (2003).

Although many people have personally experienced the phenomena I discuss, within institutional science there is a taboo against research on telepathy and other unexplained abilities, and organized skeptical groups see it as their mission to debunk any claims of the paranormal. As a result of my research, I have repeatedly come into conflict with representatives of these organizations and with professional media skeptics. In the Appendix, I summarize the main controversies in which I have been engaged. Again and again, I have found that most of my skeptical opponents are not only ignorant of the evidence but also do not want to know about it. Their minds are closed. But, as I argue in this book, science is not a dogmatic belief system but a method of inquiry. Only by investigating what we do not understand can we learn more.

This research project was initially funded by the late Ben Webster of Toronto, Canada, and has been much helped by grants from the Lifebridge Foundation in New York; the Institute of Noetic Sciences in Sausalito, California; Evelyn Hancock of Old Greenwich, Connecticut; the Ross Institute of New York; the Bial Foundation in Portugal; the Watson Family Foundation; Addison Fischer of Naples, Florida; the Planet Heritage Foundation; and the Perrott-Warrick Fund, administered by Trinity College, Cambridge University. I have also had the benefit of organizational support in the United States from the Institute of Noetic Sciences, in the German-speaking countries from the Schweisfurth Foundation in Munich, and in Britain from the Scientific and Medical Network. I am very grateful for all this generosity and encouragement.

I owe much to my research associates, Jane Turney in London, Susanne Seiler in Zurich, Switzerland, and David Brown in Santa Cruz, California. They have helped me in many ways: in carrying out surveys, in interviewing people, in doing experiments, and in collecting data. All have helped to build up a large computerized database on the perceptiveness of pets. I am also grateful to Anna Rigano and Dr. Amanda Jacks for their help with research; to Matthew Clapp for setting up my website (www.sheldrake.org) when he was an undergraduate at the University of Georgia; to John Caton for serving as my webmaster since 2002; to Helmut Lasarcyk for his labor of love in translating hundreds of reports from the German-speaking countries and adding them to our database, as well as for running my German website; and to Jan van Bolhuis for his help and advice with statistical analyses. Above all I thank my research assistant, Pam Smart, who has helped me in many ways for sixteen years and who has had the primary responsibility for maintaining and adding to my database.

Many discussions, comments, suggestions, and criticisms, as well as much practical assistance, have helped me in my research and in the writing of this book, as well as in the further research included in this new edition. In particular I thank Ralph Abraham, Shirley Barry, Patrick Bateson, John Beloff, John Brockman, Bernard Carr, Christopher Carter, Ted Dace, Sigrid Detschey, Lindy Dufferin and Ava, Sally Rhine Feather,

Peter Fenwick, David Fontana, Matthew Fox, Winston Franklin, Robert Freeman, the late Edward Goldsmith, Franz-Theo Gottwald, the late Willis Harman, Rupert Hitzig, Nicholas Humphrey, Tom Hurley, Francis Huxley, the late Montague Keene, Theodore Itten, David Lorimer, Betty Markwick, Katinka Matson, Robert Matthews, the late Terence McKenna, the late John Michell, Michael Morgan, Aimée Morgana, the late Robert Morris, the late Brendan O'Reagan, Charles Overby, Erik Pigani, Guy Lyon Playfair, Anthony Podberscek, my wife Jill Purce, Dean Radin, Anthony Ramsay, John Roche, the late Miriam Rothschild, Marilyn Schlitz, Merlin and Cosmo Sheldrake, Paul Sieveking, Martin Speich, Dennis Stillings, Harris Stone, Arnaud de St. Simon, James Trifone, Dennis Turner, Barbara Valacore, Varena Walterspiel, Ian and Victoria Watson, and Sandra Wright.

In my appeals for information I have been helped by many newspapers and magazines in Europe and North America and by a variety of TV and radio programs. I thank all those who made this possible.

I also thank all those who have given me their comments and suggestions on various drafts of this book: Letty Beyer, David Brown, Ann Docherty, Karl-Heinz Loske, Anthony Podberscek, Jill Purce, Janis Rozé, Merlin Sheldrake, Pam Smart, Mary Stewart, Peggy Taylor, and Jane Turney. I have been fortunate in having such sympathetic and constructive editors in Steve Ross and Kristin Kiser in New York and Susan Freestone in London.

Finally, I am grateful to Phil Starling for his permission to reproduce the photographs in Figures 2.1, 4.1, and 8.1; to Gary Taylor for Figure 2.2; and to Sydney King for doing the drawings and diagrams.

*London, February 2011*

# Introduction

When the telephone rings in the household of a noted professor at the University of California in Berkeley, his wife knows when her husband is on the other end of the line. How? Whiskins, the family's silver tabby cat rushes to the telephone and paws at the receiver. "Many times he succeeds in taking it off the hook and makes appreciative meows that are clearly audible to my husband at the other end," she says. "If someone else telephones, Whiskins takes no notice."

Kate Laufer, a midwife and social worker in Solbergmoen, Norway, works at odd hours and returns home at unexpected times, but whenever her husband, Walter, is home, he greets her with a hot cup of freshly brewed tea. What accounts for her husband's uncanny timing? The family dog, Tiki the terrier. "When Tiki rushes to the window and stands on the windowsill," says Walter Laufer, "I know that my wife is on her way home."

Julia Orr thought her horses had settled happily into their new paddock when she moved from Skirmett, Buckinghamshire, to a farm nine miles away. But Badger, a twenty-four-year-old Welsh Cob, and twenty-two-year-old Tango were merely biding their time. One night six weeks later, when a storm blew open the gate of their field, they took their chance. At dawn they were waiting patiently at the gate of Mrs. Orr's old home. They had found their own way back on unfamiliar roads and tracks, leaving telltale hoofprints on the shoulder of the road and in flower beds as they went.

On October 17, 1989, Tirzah Meek of Santa Cruz, California, saw her cat run up into the attic and hide, which she had never done before. She seemed terrified and refused to come down. Three hours later, the Loma Prieta earthquake struck, devastating the center of Santa Cruz.

Dogs that know when their owners are returning home, cats that answer the telephone when a person they are attached to is calling, horses that can find their way home over unfamiliar terrain, cats that anticipate earthquakes—these aspects of animal behavior suggest the existence of forms of perceptiveness that lie beyond present-day scientific understanding.

Through fifteen years of extensive research on the unexplained powers of animals, I have come to the conclusion that many of the stories told by pet owners are well founded. Some animals really do seem to have powers of perception that go beyond the known senses.

There is nothing new about the uncanny abilities of animals. People have noticed them for centuries. Millions of pet owners today have experienced them personally. But at the same time, many people feel they have to deny these abilities or trivialize them. They are ignored by institutional science. Pets are the animals we know best, but their most surprising and intriguing behavior is treated as of no real interest. Why should this be so?

One reason is a taboo against taking pets seriously.[1] This taboo is not confined to scientists but is a result of the split attitudes to animals expressed in our society as a whole. During working hours we commit ourselves to economic progress fueled by science and technology and based on the mechanistic view of life. This view, dating back to the scientific revolution of the seventeenth century, derives from René Descartes' theory of the universe as a machine. Though the metaphors have changed (from the brain as hydraulic machine in Descartes' time, to a telephone exchange a generation ago, to a computer today), life is still thought of in terms of machinery.[2] Animals and plants are seen as genetically programmed automata, and the exploitation of animals is taken for granted.

Meanwhile, back at home, we have our pets. Pets are in a different category from other animals. Pet-keeping is confined to the private, or subjective, realm. Experiences with pets have to be kept out of the real, or objective, world. There is a huge gulf between companion animals, treated as members of our families, and animals in factory farms and research laboratories. Our relationships with our pets are based on different sets of attitudes, on I-thou relationships rather than the I-it approach encouraged by science.

Whether in the laboratory or in the field, scientific investigators typically try to avoid emotional connections with the animals they are investigating. They aspire to a detached objectivity. They would therefore be unlikely to encounter the kinds of behavior that depend on the close attachment between animals and people. In this realm, animal trainers and pet owners are generally far more knowledgeable and experienced than professional researchers on animal behavior—unless they happen to be pet owners themselves.

The taboo against taking pets seriously is only one reason why the phenomena I discuss in this book have been neglected by institutional science. Another is the taboo against taking psychic, or paranormal, phenomena seriously. These phenomena are not rare or exceptional; some are very common. They are called paranormal—meaning "beyond the normal"—because they cannot be explained in conventional scientific terms; they do not fit in with the mechanistic theory of nature.

## Research with pets

The wealth of experience of animals among horse and dog trainers, veterinarians, and pet owners is generally dismissed as anecdotal. This happens so often that I looked up the origin of this word to find out what it means. It comes from the Greek root *anekdotos,* meaning "not published." An anecdote is an unpublished story.

Some fields of research—for example, medicine—rely heavily on anecdotes, but when they are published they literally cease to be anecdotes; they are promoted to the rank of case histories.

In the research described in this book, I have used three complementary approaches. First, my associates and I have interviewed hundreds of people who are experienced in dealing with animals, including dog trainers, veterinarians, blind people with guide dogs, zookeepers, kennel proprietors, and people who work with horses. I have also appealed through specialist magazines and through the general media and the

Internet for information from pet owners, and have collected more than 4,500 accounts of specific kinds of animal behavior that suggest unusual perceptiveness. I have found that many people have had very similar experiences with their animals. And when so many people's accounts point independently to consistent and repeatable patterns, anecdotes are transformed into natural history. At the very least, this is a natural history of what people *believe* about their animals.

Second, I have organized formal surveys in Britain and the United States involving random samples of households in order to quantify the frequency of the various kinds of perceptiveness shown by companion animals.

Third, I have explored the question of whether people's beliefs about their animals are well founded or not by means of experimental investigations.

One of my favorite books in biology is *The Variation of Animals and Plants under Domestication* by Charles Darwin, first published in 1868. It is full of information that Darwin collected from naturalists, explorers, colonial administrators, missionaries, and others with whom he corresponded all over the world. He studied publications like *Poultry Chronicle* and the *Gooseberry Grower's Register.* He grew fifty-four varieties of gooseberry. He drew on the experience of cat and rabbit fanciers, horse and dog breeders, beekeepers, farmers, horticulturalists, and other people experienced with animals and plants. He joined two of the London pigeon clubs, kept all the breeds he could procure, and visited leading fanciers to see their birds.

The effects of selective breeding in domesticated animals and plants, observed with such attention by practical men and women, gave Darwin his strongest evidence for the power of selection, an essential ingredient in his theory of evolution by natural selection.

Since the time of Darwin, science has increasingly cut itself off from the rich experience of people who are not professional scientists. There are still millions with practical experience of pigeons, dogs, cats, horses, parrots, bees, and other animals, and of apple trees, roses, orchids, beans, asparagus, and other plants. There are still tens of thousands of amateur naturalists. But scientific research is now almost entirely confined to universities and research institutes and carried out by professionals with Ph.D.'s. This exclusivity has seriously impoverished modern biology.

## Why hasn't this research been done before?

The investigation of the unexplained powers of animals that I describe in this book has been facilitated by modern technical devices such as computers and video cameras, but in principle most of these investigations could have been carried out a hundred or more years ago. The fact that they are only now beginning is a tribute to the strength of the taboos against such inquiries.

I believe there is much to be gained by ignoring these taboos. I also believe there is much to be gained by following a scientific approach. This is the approach I have followed myself, and which I summarize in this book. But the word "scientific" can have quite different meanings. All too often it is equated with a narrow-minded dogmatism that seeks to deny or debunk whatever does not fit in with the mechanistic view of the world. By contrast, I take "scientific" to mean a method of open-minded inquiry, paying attention to evidence and testing possible explanations by means of experiment. The path of investigation is more in the spirit of science than the path of denial. And it is certainly more fun.

These different scientific attitudes are illustrated by the tale of a horse called Clever Hans, which is usually used to justify the dismissal of seemingly unexplained animal powers. I draw the opposite moral from the story and see it as an example of the need to investigate rather than deny unexplained phenomena. Sooner or later anyone who takes an interest in the unexplained power of animals will be told the story of Clever Hans. This story has assumed the role of a cautionary tale for scientists.

## The tale of Clever Hans

In Berlin at the beginning of the twentieth century there was a horse named Hans, who was said to be able to perform mathematical calculations, read German, and spell out German words. He tapped out answers with his hoof. His trainer, Herr von Osten, a former mathematics teacher, was convinced that Hans had mental capacities thought to be confined to human beings. The horse caused a sensation, and gave many demonstrations for professors, military officers, and others.

Clever Hans's abilities were investigated by Professor C. Stumpf, director of the Psychological Institute of the University of Berlin, and his assistant, Otto Pfungst. They found that the horse could give the correct answers only when the questioner knew the answer himself and when Hans could see the questioner. They concluded that Hans had no mathematical abilities and he could not read German. Instead, he was reading small body movements of the questioner, and these told him when he had tapped with his hoof the right number of times.

This tale of Clever Hans has been used ever since to justify the dismissal of unexplained abilities of animals, attributing them to subtle cues rather than to any mysterious powers the animal might have. In short, this story has been used to inhibit research, to prevent inquiry rather than to stimulate it. But to draw this moral from the tale of Clever Hans does not do justice to the investigations of Stumpf and Pfungst. They investigated a controversial claim rather than dismissing it, and they were brave to do so, because their conclusions went against the beliefs of many of their colleagues.

Clever Hans's abilities were controversial not because they were supposed to involve psychic powers but rather because they were supposed to show that animals could think. Many scientists, especially Darwinians, were happy to believe that Clever Hans really could do arithmetic and understand German. They liked the idea that animals were capable of rational thought because this undermined the conventional belief that the human intellect was unique. They preferred the idea of gradual evolution, of differences of degree rather than differences of kind between humans and nonhuman animals.

Conversely, traditionalists were very skeptical about Clever Hans because they thought that higher mental faculties were confined to man. Stumpf and Pfungst's findings supported the traditionalists and were unpopular with "disappointed Darwinians who expressed fear lest ecclesiastical and reactionary points of view should derive favorable material from the conclusions."[3]

Although biologists sometimes talk about the "Clever Hans effect" as if it were a reason for dismissing any unexplained abilities in animals, the effect is quite specific. It depends on body language, which in horses is an important element in their communication with one another, as it is in many other species. If an animal can respond to a human being when that person is out of sight, this is not an example of the Clever Hans effect, but requires some other explanation.

In the course of research on the unexplained powers of domestic animals, I have found that most animal trainers and pet owners are well aware of the importance of body language. But in any case, many of the phenomena I discuss here, such as the apparent ability of animals to know when their owners are coming home, cannot be explained in terms of the Clever Hans effect. An animal cannot read the body language of a person many miles away.

## Three kinds of unexplained perceptiveness

In this book I discuss three major categories of unexplained perceptiveness by animals: telepathy, the sense of direction, and premonitions.

**Telepathy.** I start with the ability of some dogs and other animals to know when their owners are coming home. In many cases an animal's anticipation of a person's return cannot be explained in terms of routine, clues from people at home, or the sound of a familiar car approaching. In videotaped experiments, dogs can still anticipate their owners' return at randomly chosen times, even when the owners are traveling by taxi or some other unfamiliar vehicle. Somehow people telepathically communicate their intention to return home.

Some companion animals also respond telepathically to a variety of other human intentions and react to silent calls and commands. Some know when a particular person is on the telephone. Some react when their owner is in distress or dying in a distant place.

I suggest that telepathic communication depends on bonds between people and animals—bonds that are not mere metaphors but actual connections. They are connected through fields called morphic fields. I introduce these fields in Chapter 1, in which I also discuss the evolution of the bonds between humans and animals.

**The sense of direction.** Homing pigeons can find their way back to their cote over hundreds of miles of unfamiliar terrain. Migrating European swallows travel thousands of miles to their feeding grounds in Africa and, in the spring, return to their native place, even to the very same building where they nested before. Their ability to navigate toward distant destinations is still unexplained and cannot be accounted for in terms of smell or any of the other known senses, or even a compass sense.

Some dogs, cats, horses, and other domesticated animals also have a good sense of direction and find their way home from unfamiliar places many miles away. Animals seem to be drawn toward their desired destination as if by an invisible elastic band that attaches them to that place. These connections may be explained in terms of morphic fields.

Sometimes animals return not to places but to people. Some dog owners who have gone away and left their pets behind are found by the animal in distant places the animal has never been to before. Tracking the person by smell may explain some cases when the distances are short, but in others the only feasible explanation seems to be an invisible connection between the animals and the people to whom they are bonded. Again, this could be compared to a stretched elastic band, which I attribute to the morphic field linking animal to owner.

**Premonitions.** Some premonitions may be explicable in terms of physical stimuli—animals that become disturbed before earthquakes, for example, may be reacting to subtle electrical changes. Dogs that alert their epileptic owners to an impending fit may notice subtle muscular tremors or unusual odors. But other premonitions seem to result from a mysterious foresight that challenges our assumptions about the separation of past, present, and future.

Telepathy, the sense of direction, and precognition are examples of what some people call extrasensory perception, or ESP. Others attribute them to a sixth or seventh sense. Others call them paranormal. Still others call them psychic. All these terms point beyond the limits of established science.

"Extrasensory perception" literally means perception beyond or outside the senses. At first sight the term "sixth sense" appears to mean the opposite because it implies a perceptiveness within the senses, although by another kind of sense not yet recognized by science. This conflict vanishes if "extrasensory" is taken to mean "outside the *known* senses."

Neither the term "extrasensory perception" nor the term "sixth sense" suggests what these experiences are or how they work. The terms merely tell us what the events are not: they are not explicable in terms of the known senses.

All three types of perceptiveness—telepathy, the sense of direction, and premonitions—seem better developed in nonhuman species than they are in people, but they do occur in the human realm too. Human psychic powers seem more natural, more biological, when they are seen in the light of animal behavior. Much that appears to be paranormal at present looks normal when we expand our notion of normality.

Science can advance only by going beyond its current limits. In this book I hope to show that it is possible to investigate animals' unexplained abilities scientifically in ways that are neither invasive nor cruel. I also suggest a variety of ways in which animal owners and students could make major contributions to this new field of inquiry.

We have a great deal to learn from our companion animals. They have much to teach us about animal nature—and about our own.

# Part I

# Human-Animal Bonds

*Chapter 1*

# The Domestication of Animals

Many people love their pets and are loved by them. In this chapter I explore the evolution and the nature of human-animal bonds.

But first it is important to recognize that emotional bonds between people and animals are the exception rather than the rule. For every well-loved cat or dog, hundreds of domesticated animals are confined to barren environments in intensive farming systems and research laboratories. In many Third World countries beasts of burden are often treated brutally. And traditional societies are not usually subscribers to modern ideals of animal welfare. Eskimos, for example, tend to treat their huskies harshly.

But in spite of all this exploitation, abuse, and neglect, many people form bonds with animals from childhood onward. Young children are commonly given teddy bears or other toy animals, and they like hearing stories about animals. Above all, most like keeping actual animals. The majority of pets live in households with children.[1]

Hearing tales about frightening animals, including the wolf in "Little Red Riding Hood," and forming relationships with friendly ones seems to be a normal and fundamental aspect of human nature. Indeed our nature has been shaped throughout its evolutionary history by our interactions with animals, and all human cultures are enriched by songs, dances, rituals, myths, and stories about them.

## The evolution of human-animal bonds

The earliest named hominid species, known from fossil remains, are *Australopithecus ramidus* and *Australopithecus anamensis*, dating back over 4 million years. The first stone tools were used about 2½ million years ago, and signs of meat eating appear about a million years later, around the time that *Homo erectus* spread out of Africa into Eurasia (Figure 1.1). The use of fire may have begun around 700,000 years ago. Modern humans originated in Africa about 150,000 years ago. The first cave paintings, including many of animals, appeared about 30,000 years ago. The agricultural revolution began about 10,000 years ago, and the first civilizations and written scripts about 5,000 years ago.[2]

Our ancestors lived as gatherers and hunters, with gathering far more important than hunting. The old image of man the hunter striding confidently out onto the African veldt is a myth. Only a small proportion of the food eaten by today's hunter-gatherers comes from animals hunted by the men; most comes from gathering done mainly by women. The exceptions are the hunter-gatherers of the plant-poor Arctic regions.[3] Hominids and early *Homo sapiens* obtained small amounts of meat more by scavenging the kills left by more effective predators like big cats than by hunting for themselves.[4] Big game hunting, as opposed to scavenging, may date back only some 70,000 to 90,000 years.

In hunter-gatherer cultures, human beings do not see themselves as separate from other animals but as intimately interconnected.[5] The specialists in communication with the nonhuman world are shamans, and through their guardian spirits or power animals, shamans connect themselves with the powers of animals. There is a mysterious solidarity between people and animals. Shamans experience themselves as being guided by animals or as changing into animals, understanding their language, and sharing in their prescience and occult powers.[6]

## The earliest domesticated dogs

The first animals to be domesticated were dogs. Their ancestors, wolves, hunted in packs, just as men hunted, and from an early stage dogs

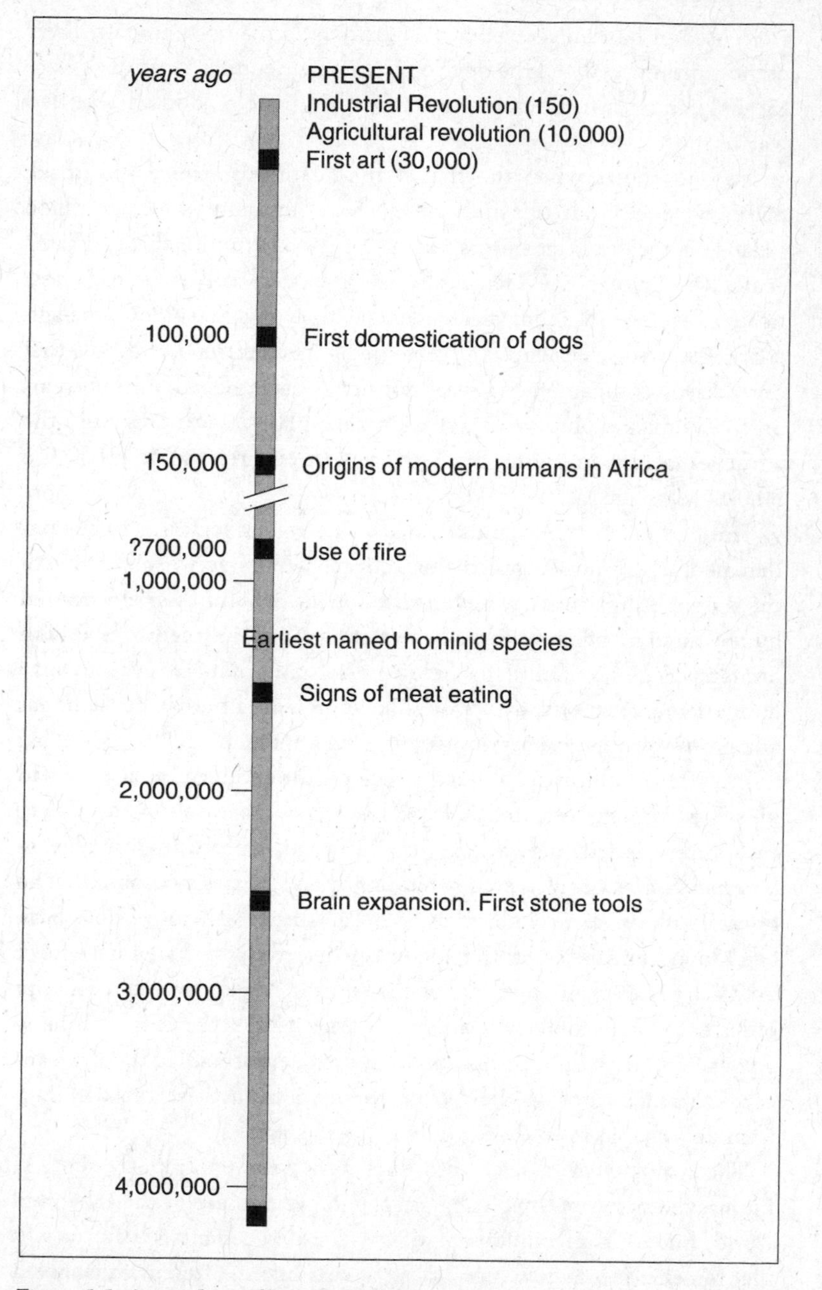


*Figure 1.1 A time line of human evolution.*

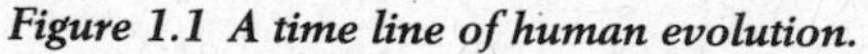

were used in hunting as well as for guarding human settlements. Their domestication predated the development of agriculture,[7] and dogs were the only animals to be domesticated before people adopted a settled way of life.[8]

No one knows when the first domestication of wolves occurred. Some evidence from the study of DNA in dogs and wolves points to a date for the first transformation of wolf to dog more than 100,000 years ago. This DNA evidence also suggests that wolves were domesticated several times, not just once, and that dogs have continued to crossbreed with wild wolves.[9] If this theory is confirmed, it means that our ancient companionship with dogs may have played an important part in human evolution. Dogs could have played a major role in the advances in human hunting techniques that occurred some 70,000 to 90,000 years ago.

The Australian veterinarian David Paxton goes so far as to suggest that people did not so much domesticate wolves as wolves domesticated people. Wolves may have started living around the periphery of human settlements as a kind of infestation. Some learned to live with human beings in a mutually helpful way and gradually evolved into dogs. At the very least, they would have protected human settlements, and given warnings by barking at anything approaching.[10]

In 2009 an international team of scientists announced that they had identified the earliest archaeological evidence of a dog in the Goyet cave in Belgium, dating back 31,700 years ago. It probably resembled a Siberian Husky but was somewhat larger, and it subsisted on a diet of horse, musk ox, and reindeer.[11] Other Paleolithic remains of dogs have been found in Russia and the Ukraine, where they may well have been used for the tracking, hunting, or transport of large game animals. Still other early archaeological evidence, a track of footprints from a large dog walking with a child, was found in the deepest part of the Chauvet cave in France. Soot on the roof of the cave, from the torch the child was carrying, has been dated to 26,000 years ago.

The wolves that evolved into dogs have been enormously successful in evolutionary terms. They are found everywhere in the inhabited world, hundreds of millions of them. The descendants of the wolves that remained wolves are now sparsely distributed, often in endangered populations.

*Figure 1.2 Breeds of Egyptian dogs, from the tombs at Beni Hassan (2200–2000 B.C.) (after Ash, 1927).*

The domestication of dogs long predated the domestication of other animals. Indeed dogs may have played an essential part in the domestication of other species, both through their ability to herd animals such as sheep and also by helping to protect flocks against predators.

Some breeds of dogs are very old. By the time of ancient Egypt, there were already several distinct breeds: dogs of the Greyhound or Saluki type, a Mastiff type, a Basenji type, a Pointer type, and a small terrier-like Maltese type (Figure 1.2).[12] Dogs were venerated in ancient Egypt. Some were even embalmed, and in every town a graveyard was devoted entirely to dog burials. The god of the dead was the dog- or jackal-headed Anubis.

In today's world, there are great variations from culture to culture in the way dogs are treated. In the Arab world, they are generally abhorred, partly because of large populations of stray or feral dogs, a source of dangerous diseases such as rabies. Even so, individual hunting dogs are admired and pampered. In other places, as in parts of China, Burma, Indonesia, and Polynesia, dogs are slaughtered for human food and are not usually well regarded.[13] But in most cultures, especially where dogs are used for hunting or herding, or kept for no utilitarian reason, they are generally treated affectionately.[14]

Although the domestication of dogs happened so long ago that we will never know the details, a twentieth-century study in Russia with silver foxes showed that quite rapid changes can occur under conditions of selective breeding. From the 1950s onward, tame foxes were selected as the parents of the next generation, and after forty generations, the Russians succeeded in producing a breed of silver foxes that are docile, friendly, and as skilled as dogs in communicating with people.[15] The tame foxes also look different from their wild ancestors, with broader heads and juvenile characteristics.[16] Some of these animals are now being sold as pets.

## The domestication of other species

Francis Galton, Charles Darwin's cousin, was a pioneer of modern thinking about domestication. He pointed out that relatively few species are suitable. Species capable of being domesticated have to meet certain conditions: They should be hardy and able to survive with little care and attention. They should have an inherent fondness for humans. They should be comfort-loving and useful. They should breed freely, and they should be gregarious and hence easy to control in groups.

Sheep, goats, cattle, horses, pigs, hens, ducks, and geese all meet these criteria. Other species, such as deer and zebras, although gregarious, do not, and despite many attempts at domestication they remain too wild to manage with ease.[17]

Cats are the only domesticated animals that are not gregarious, but because of their territorial and comfort-loving nature, they form symbiotic relationships with people while preserving something of their independence as solitary hunters. They revert with relative ease to a free-living, feral existence.[18]

Cats were domesticated much more recently than dogs, probably no more than 10,000 years ago. The oldest archaeological evidence of cats comes from Crete, about 9,500 years ago, and cat remains from Jericho have also been dated to 8,700 years ago.[19] The first records of cats are from ancient Egypt, where they were treated as sacred, and it was forbidden to kill them. Some 3,600 years ago, house cats were depicted in Egyptian tomb paintings. They were mummified in such enormous numbers that at the beginning of the twentieth century cat mummies were excavated by the ton, ground up, and sold as fertilizer.[20]

Horses were also domesticated relatively recently, probably about 5,000 years ago in the region around Turkestan. They may first have been used as draft animals. The first record of a horse being ridden is from Egypt, around 1500 B.C.[21] Horses soon became important in war and in hunting, where they were more like comrades than slaves.

In early civilizations, although domesticated animals were exploited for human use, there was still a pervasive sense of human-animal connectedness. Many animals were regarded as sacred, just as cows, elephants, and monkeys still are in India today. Many of the gods and goddesses were believed to take on animal forms or to have animal helpers.

At first sight, there is little trace of this solidarity with the animal kingdom in industrial societies. Beasts of burden have been replaced by machines; horses, donkeys, mules, and bullocks are no longer our daily companions. The peasant's intimate familiarity with animals has been replaced by modern agribusiness, with animals kept on factory farms and in industrial-scale feed lots.

Nevertheless, in our private lives, the ancient affinity with other animals remains. There are many bird-watchers, naturalists, and amateur wildlife photographers. Wildlife films are perennial favorites on television, as are stories about animals, especially about dogs like Lassie[22] and the Austrian detective dog, Kommisar Rex. But it is principally and most intimately through the keeping of pets that these bonds are maintained. Even though most people in modern cities no longer need cats for mousing or dogs for herding or hunting, these animals are still kept in the millions, together with a host of other creatures that play no utilitarian role: ponies, parrots, budgerigars, rabbits, guinea pigs, gerbils, hamsters, goldfish, lizards, stick insects, and many other pets.

Most of us seem to need animals as part of our lives; our human nature is bound up with animal nature. Isolated from it, we are diminished. We lose a part of our heritage.

## The keeping of pets

All over the world people keep pets. As Francis Galton noted in 1865: "It is a fact familiar to all travelers that savages frequently capture young animals of various kinds, and rear them as favorites, and sell or present them as curiosities."[23]

Galton suggested that the principal way in which many species were first tamed was through this kind of pet-keeping, together with the keeping of sacred animals and the maintaining of menageries by chiefs and kings. In some cases, these animals became domesticated if they met the necessary criteria (summarized on page 18). I like Galton's suggestion that pet-keeping preceded domestication, and I find it very plausible. And if wolves first became camp followers and then evolved into dogs, Galton's theory suggests a simple way in which this process could have been speeded up, through people adopting cubs or puppies as pets.

In ancient Egypt and many other parts of the world, in addition to the larger dogs used for hunting, guarding, and herding, there were smaller breeds that seem to have lived in houses as pets. Ancient Greeks and Romans also kept house pets (Figure 1.3). Indeed small dogs were found all over the ancient world and are the ancestors of many pet dogs of today. In Tibet and China it was customary to keep both guard dogs and house dogs; guard dogs were big and fierce and lived outside, while the small dogs lived indoors in houses and monasteries.[24]

Pet-keeping, unlike the keeping of working animals, was something of a luxury in ancient times. Far more people are affluent today, and more keep pets. And pets living indoors as companions often become more intimately connected to their human family than animals living in a farmyard, barn, or kennel. In industrialized countries like France, Britain, and the United States, the majority of households contain at least one companion animal. And over recent decades, as urbanization and prosperity have increased, even more households have kept pets. In the United Kingdom, for example, between 1965 and 2010 the total number of dogs rose from 4.7 to 8 million, and of cats from 4.1 to 8 million.[25]

The animal-keeping habits of different nations probably play a large

***Figure 1.3 Small pet dogs in ancient Greece (after Keller, 1913).***

part in the forming of national character. But this is an area where there has been almost no research; we have only bare statistics. The highest percentages of households with dogs are in Poland and the United States, with France, Belgium, and Ireland next. Some of the lowest levels of dog and cat ownership are in Germany. In most countries, more households contain dogs than cats, but in some, notably Switzerland and Austria, there is a striking preference for cats as house pets.

In recent years, some changes have occurred in the pattern of pet ownership. In France the percentage of households with both dogs and cats has declined. In Germany and Switzerland the number of dogs

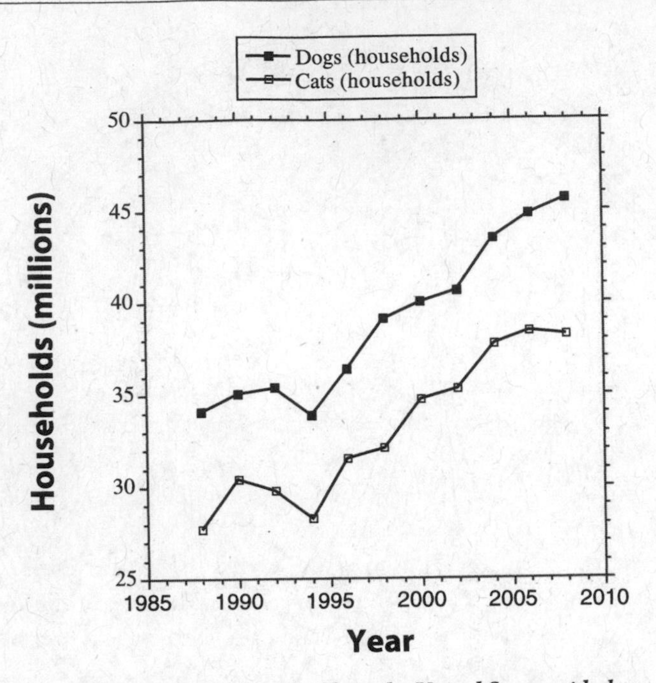


*Figure 1.4 The number of households in the United States with dogs and cats, from 1988 to 2008. (Source: American Veterinary Medical Association, 2010)*

remained more or less the same, while the number of cats increased, but there were still striking national differences. In 2008 the percentages of households with dogs were France, 24; United Kingdom, 22; Germany, 14; and Switzerland, 12. The percentages with cats were United Kingdom, 28; France, 27; Switzerland, 25; and Germany, 16. In the United States there has been a steady increase in the number of both dogs and cats (Figure 1.4), but consistently more households had dogs than cats. In 2008 39 percent of U.S. households had dogs and 33 percent had cats, higher than any country in Western Europe. And of course the numbers of dogs and cats are higher than the numbers of households because many people have more than one cat or dog. In 2008 U.S. households had an average of 1.7 dogs and 2.2 cats.[26]

Although dogs and cats are the most popular pets, many households also keep birds, reptiles, fish, and small mammals like rabbits, guinea pigs, hamsters, and ferrets. In 2008 62 percent of U.S. households had at least one pet.

## Social bonds between animals

Most domesticated animals were originally social, as Francis Galton pointed out. They also tend to be animals with dominance hierarchies, which made it easier for human beings to control them. Even cats, although independent and solitary in their hunting habits, grow up with close social relationships between mothers and their offspring.

The original social nature of domesticated animals reveals itself when they run wild. Charles Darwin, in *The Variation of Animals and Plants under Domestication*, was particularly interested in this reversion of domesticated animals to their ancestral habits.[27]

In general feral animals live in groups similar to those of their wild progenitors. Feral horses, for example, usually live in groups of around five, and so do their wild relatives.[28] Feral dogs live in packs and build dens, as do their wolf ancestors.[29]

Social animals are linked to other members of their group by invisible bonds. The same is true of human social bonds. Our domesticated animals are by nature social, and so are we. The bonds between people and animals are a kind of hybrid between the bonds that animals form with each other and those that people form with each other.

One difficulty in understanding the nature of these animal-human bonds is that we understand so little of human-human and animal-animal bonds. We know that invisible emotional connections exist between members of a family, and we know that these can persist over time and keep people linked together even when they are continents apart. We know animals have social groups and that somehow the group as a whole is linked together so that it can function as if it were a superorganism, as I discuss in Chapter 9. This is most clearly observable in the social insects, like the ants, termites, bees, and wasps. It is plainly visible in a flock of birds turning and banking practically simultaneously, with none of them bumping into each other. And so it is with a school of fish swimming in close formation, but changing direction at any time, and responding rapidly to the approach of a predator.

## The nature of social bonds

There are many kinds of social bonds within species, like those between a mother cat and her kittens, a bee and the other members of the hive, a starling in a flock, a wolf and its pack, and a great variety of human social bonds. Then there are social bonds *between* species, like those between pets and their owners.

All of these bonds connect the members of a group together and influence the way they relate. I propose that these bonds are not just metaphorical but real, literal connections. They continue to link individuals together even when they are separated beyond the range of sensory communication. These connections at a distance could be channels for telepathy.

Bonds between animals exist within a *social field*. Like the known fields of physics, social fields connect things at a distance, but they differ from the known fields of physics in that they evolve and contain a kind of memory. I have suggested in my book *The Presence of the Past* that social fields are an example of a class of fields called morphic fields.[30]

Morphic fields hold together and coordinate the parts of a system in space and time, and contain a memory from previous similar systems. Human social groups such as tribes and families inherit through their morphic fields a kind of collective memory. The habits, beliefs, and customs of the ancestors influence the behavior of the present, both consciously and unconsciously. We all tune in to collective memories, similar to the collective unconscious proposed by the psychologist Carl Jung.

Termite colonies, schools of fish, flocks of birds, herds, packs, and other animal groups are also held together and structured by morphic fields, and these fields are all shaped by their own kinds of collective memory.

Individual animals are linked together through the social fields of their group. They follow habitual patterns of relationships, repeated over the generations. Instincts are like collective habits of the species, or of the breed, shaped by experience through many generations, and subjected to the rigors of natural selection. This view of instincts as the inherited effects of habit and experience is close to the thinking of Charles Darwin, most clearly expressed in *The Variation of Animals and*

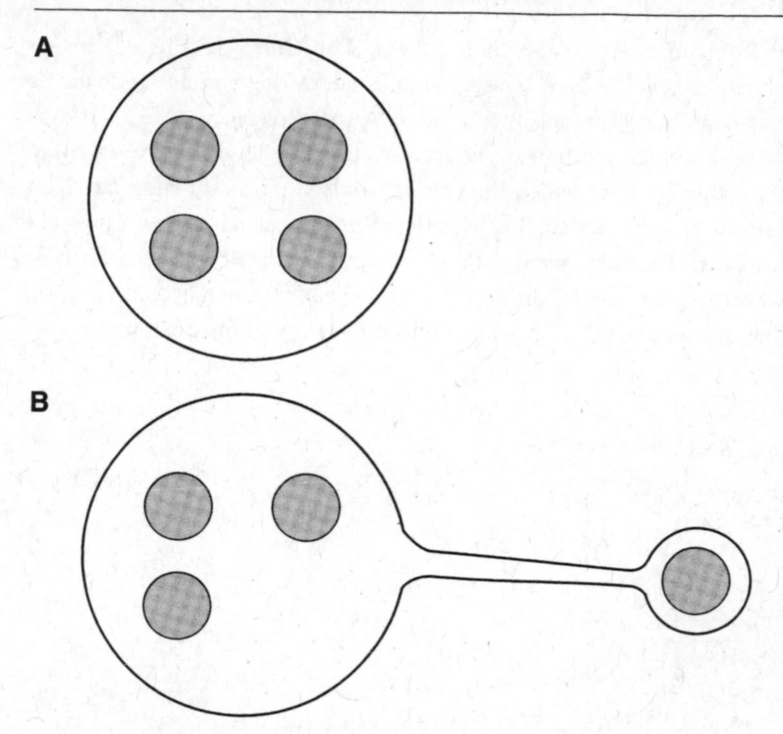


***Figure 1.5** A diagrammatic representation of the morphic field of a social group (A), illustrating the way the field stretches out and still connects an individual with other members of the group when they are far apart (B).*

*Plants under Domestication* and of central importance in *The Origin of Species.*[31]

The process by which this memory is transferred from past to present is called morphic resonance, involving an influence of like upon like across space and time.[32] (I discuss the nature of morphic fields and morphic resonance in more detail in Chapter 9.)

Morphic fields link together the members of a social group. A field embraces all the members of the group within itself (Figure 1.5A). A member who goes to a distant place still remains connected to the rest of the group through this social field, which is elastic (Figure 1.5B).

Morphic fields would permit a range of telepathic influences to pass from animal to animal within a social group, or from person to person, or from person to companion animal. The ability of these fields to stretch out like invisible elastic bands enables them to act as channels for telepathic communication, even over great distances.[33]

At this stage, it is not necessary to grasp the details of the morphic field hypothesis, of which I have given only the briefest summary. The important point is that this hypothesis makes telepathy seem possible and even likely. But given that it is theoretically possible, does it actually occur? On the basis of the available evidence, discussed in the following chapters, I conclude that telepathy is indeed a real phenomenon.

Part II

# Animals That Know When Their People Are Coming Home

## Chapter 2

# Dogs

The most convincing evidence for telepathy between people and animals comes from the study of dogs that know when their owners are coming home. This anticipatory behavior is common. Many dog owners simply take it for granted without reflecting on its wider implications.

When Peter Edwards arrives home at his farm in Wickford, Essex, his Irish Setters are nearly always at the gate to greet him. Yvette, his wife, says they often wait for him for ten to twenty minutes before he arrives and well before he turns off the road into his drive. She had taken this behavior for granted for years, simply thinking, "Peter's coming home, the dogs have gone to the gate."

Yet after reading in the *Sunday Telegraph* about my research on dogs that know when their owners are coming home, Yvette began to wonder how the setters knew when Peter was coming. He worked irregular hours in London and did not usually let her know when to expect him. The dogs responded regardless of which way the wind was blowing or what vehicle he was driving.

The Irish Setters' ability to predict Peter's return is typical of many dogs. In response to my appeals for information and my interviews with kennel proprietors, animal trainers, and pet owners in Europe, North America, and Australia, I have collected more than one thousand reports of dogs that know when their owners are coming home. Some wait at a door or window for ten minutes or more in advance of their owners' return from work, school, shopping, or other excursions. Others go out and meet their owners in the street or at a bus stop. Some dogs do this on an almost daily basis, others only when their owners are returning from a holiday or other protracted absence, sometimes showing signs of excitement hours or even days in advance. While some scientists are quick to attribute this phenomenon to routine or to the canine's sharp sense of smell and hearing, you will soon discover that in case after case no such simple explanations suffice.

The context for this anticipatory behavior is the way that many dogs welcome their owners home with great enthusiasm. Unless they are well disciplined they try to jump up and lick their owners' faces, just as puppies greet their parents, with their tails wagging so vigorously that the whole hindquarters become part of the movement.

Wolf greetings are similar. When cubs are weaned, they start soliciting food from their returning parents or other members of the pack. When the adult approaches with food in its mouth, they crowd excitedly around, wag their tails, assume gestures of submission, and jump up to lick the corners of the adult's mouth. In adult wolves, the same kinds of behavior develop into ritualized greetings. Most attention is directed to the highest-ranking animals.[1]

Thus the greeting behavior that dogs display toward their owners has had a long evolution, going back to the wolves from which our domestic dogs are descended. But many dogs not only greet their owners on arrival but actually anticipate their arrival, seeming to know when they are on the way home even when they are many miles away.

## Could it be a matter of routine?

When people return at the same time every day, their dogs' behavior could simply be a matter of routine. Teresa Preston, of Suffolk, Virginia, assumed this was the case when she noticed that the family dog, Jackson, waited for her children's arrival from school each day. But she had to think again when she realized that Jackson also anticipated the return of her husband, who arrived at unexpected times from his job as captain of a U.S. Coast Guard buoy tender stationed 20 miles away in Portsmouth:

> He would arrive home at odd hours. When the ship had come into port, Jackson would get excited, go to the door, and want out. Most of the time he would go and sit at the end of the sidewalk, stationing himself to look in the direction he knew the car would travel. He got so good at this, I couldn't help but notice, and sometimes would use Jackson's warning to freshen my hair and makeup before my husband arrived! If I was cooking dinner and was at the point of deciding how many portions to cook or places to set for the meal, I would use his prediction and add accordingly.

Or perhaps the dogs are picking up clues of anticipation from the person waiting at home. In some cases people telephone to say they are coming, and the emotional state of the person at home might then change, causing him or her to give the dog clues through body language or in other ways. But some dogs anticipate a return even when those at home have no idea when the person will arrive. I have received numerous accounts from the families of lawyers, taxi drivers, military personnel, journalists, midwives, and other people who do not work fixed hours, who say that it is the dog who tells *them* when their family member is on the way home. For example, Rebecca Kavich, who lives in Australia, said:

> I could always tell when my husband was on his way home because his dog would tell me! His business is approximately a ten-minute drive from home, and about fifteen minutes before he'd arrive our Husky, Zero, would start to get agitated and excited. He'd follow me around and run back and forth to the front door waiting and looking . . . and waiting. Tony would come home at different times each day depending on his schedule, but it seems Zero sensed when Tony was closing up and heading home.

In Manhattan, the West family's Irish nanny benefited from a similar canine early warning system, in the form of a Kerry Blue Terrier. General

Charles West was stationed on Governors Island, in New York Harbor, and his wife worked as a vice president at Time Inc. In General West's words:

> We lived on the fourth floor of an apartment building, and each of us arrived home at varying times from varying directions. Neither the nanny nor our small son knew when we were coming home, but ten to fifteen minutes before our arrival Kerry would get greatly excited, run to a front window and stand looking into the street, whining joyously, with her tail going like mad. Nanny always knew that one of us was about to arrive, and she always laughed that it was a great warning to clean up the child before the parent got there. And this was not an occasional happening. It went on day after day, week after week for years.

Some people have done simple tests to find out whether their dogs are reacting at a routine time. When David Speck was living with his partner in Battery Park, New York, his Tibetan Terrier, Sophie, became very excited when he was coming home from the office: "I randomized my arrival time sometimes by several hours, but time and again Sophie showed signs of excitement at least ten minutes before my arrival home. Typically she would scamper through the laundry hamper and celebrate with a dirty sock of mine and greet me at the door with this prize."

No doubt some dogs are accustomed to waiting for the return of their owner at routine times, but most people do not regard this as particularly remarkable. In most of the 1,133 reports I have received, as in these examples, the dog's behavior is not explicable simply in terms of routine.

## Could dogs smell their owners approaching?

Most dogs have a much better sense of smell than we do, and it is likely that they could smell their owners, or their owners' vehicles, from farther away than a person could. But just how far?

Dogs normally use their sense of smell for tracking, sniffing the ground, and following a trail. But to smell someone returning home, they would have to sniff the air. Assuming that the wind is blowing in the right direction and they are outdoors, or indoors with the windows open, over what range might they be able to smell an approaching person or car?

The best estimates I have been able to obtain suggest that this distance is considerably less than a mile, even with the most sensitive of all breeds, the Bloodhound. Malcolm Fish, of the Essex Police Dog Section, conducted trials of Bloodhounds for the British government's Home Office to find out if they would be more suitable for some kinds of police work than German Shepherds, at that time the standard breed. He says that if somebody is hiding in a hedge, a Bloodhound up to half a mile downwind can sometimes pick up the scent of that person, but only if the wind is blowing in the right direction and if the person is stationary. He thinks it highly unlikely that a dog, even a Bloodhound, could smell someone traveling home from work. "If you imagine someone in a car traveling home with a smoke canister, with the windows open and the smoke blowing out, it would blow behind. Scent does not travel forward like sound. Nowadays most cars are sealed too, so there would not be much scent leaving the car, and doors of houses are sealed to keep the draft out, so I think it would be impossible for dogs to smell their owners when they are half a mile away."

Smell might help explain why some dogs react only a minute or two before their owners arrive, but many react ten minutes or more in advance, when the person is several miles away. Moreover, they do so irrespective of the wind direction, and can still do it even when the windows are shut. Their anticipation cannot reasonably be explained in terms of smell.

## Could dogs hear their owners approaching?

Most dogs have sharper hearing than we do. They can hear sounds too high-pitched for us to detect, as in dog whistles that emit sounds above the frequency range we can hear. They may also hear sounds farther away. One rough estimate is that "a dog can hear sounds four times farther away than a human can."[2] But this may be unduly generous to dogs. Celia Cox, a British vet who specializes in ear, nose, and throat surgery, has tested the hearing of thousands of dogs and estimates that their sensitivity to noise levels is similar to that of people. She doubts that they can hear their owners approaching from very far away: "People have told me that their dogs know when they are coming home even before they have turned into their road, but I think it is highly unlikely that this is purely due to hearing."

Likewise, Kevin Munro of the Hearing and Balance Centre at the University of Southampton has compared the hearing ability of people with that of dogs using a sophisticated technique called Evoked Response Audiometry.[3] He was expecting to find that dogs hear much better than people, because this is such a common belief. "When I got the results I was very surprised that their hearing, apart from being able to hear higher-pitched sounds, was in every other way similar."

But for the purpose of argument, let us assume that dogs really can hear things about four times as far away as people can. If a familiar car or person on foot is approaching your house, how far away do you hear them?

I live in London, and with all the background noise and passing cars and people, I probably hear only familiar cars or people approaching my home less than 20 yards away, and even then only when I am in one of the rooms at the front of the house with the windows open. By contrast, people in isolated parts of the countryside with little or no passing traffic might hear an approaching vehicle more than half a mile away, especially at night. But I estimate that in urban and suburban settings most people would not recognize the sounds of a familiar car or person more than a few hundred yards away, and generally much less than that. You can make your own estimate. And then you can test it with the help of your family and friends. Can you really tell that a particular car is approaching when it is that far away?

Multiply your estimate by four. Then you have a rough indication, on the most generous of assumptions, of how far away a dog could hear its owner's return. My guess would be that in urban and suburban settings this distance would be less than half a mile, even under the most favorable conditions, with the wind blowing in the right direction. With the wind blowing in other directions, the range would be much smaller. It would be smaller still if the dog was indoors with the windows closed.

All this assumes that the person is traveling on foot or in a familiar car, but what if the person is traveling in a taxi, a friend's car, or any other vehicle with which the dog is unfamiliar? In spite of the lack of familiar sounds to recognize, many owners have found that the dogs can still anticipate an arrival.

For example, when Louise Gavit, of Morrow, Georgia, sets off to come home, the family dog, BJ, goes to the door. Mrs. Gavit's husband has seen BJ do this over and over again, and by keeping note of the time, he has found that BJ usually begins to react when Louise first decides to come home and starts walking toward whatever vehicle she plans to return home in, even when she is many miles away. "My method of travel is irregular: I may use my own car, my husband's car, a truck, or any number of cars driven by strangers to BJ, or I might walk. Somehow BJ responds to my thought-action just the same. Even when he has seen my car still inside the garage he reacts."

## Returns by bus, train, and plane

The idea that dogs' reactions might be explained in terms of distant car sounds is also refuted by the fact that dogs can anticipate the arrival of owners who are traveling by bus or train. Of course, if they ride the same bus—say, a school bus—the animal might recognize characteristic sounds before the vehicle arrives. But when people travel at different times on buses or trains, there is no way the animal could tell from the sound whether the owner is on a particular bus or train.

Helen Meither, for example, commuted 15 miles by bus to work in Liverpool each day, leaving her Cairn Terrier with her family. Depending on when she finished her work, she might come home on a bus that arrived at 6:00 P.M. or on one that arrived at 8:00 P.M. "The bus stop was about a quarter of a mile away through a small wood. I never knew whether I would finish work in time to catch the earlier bus, but the dog always knew whether I was on it. If I was, he went to the door about 5:45 to 5:50 P.M., whatever the weather, and came across the wood to meet me. If I was late he did not stir until about 7:45 and met me at the later bus."

On the database there are more than sixty accounts of dogs reacting to people's arrival by bus that show that the animal somehow knows when the person is coming home in a way that cannot be explained in terms of routine, sounds, or smells. The same is true of more than fifty cases involving travel by train. Here is one example:

Carole Bartlett of Chiselhurst, Kent, leaves Sam, her Labrador-Greyhound cross, at home with her husband when she goes to the theater or visits friends in London. She returns from Charing Cross station, a twenty-five-minute train journey followed by a five-minute walk. Mr. Bartlett does not know which train she will return on; she could arrive any time from 6:00 to 11:00 P.M. "My husband says Sam comes downstairs off my bed, where he spends the day when I go out, half an hour before my return and waits at the front door." In other words, the dog begins waiting for her around the time she is starting her train journey.

In some cases, the absent person tells the person at home that he or she will take one particular train, then in fact takes another. This happened when Sheila Brown, of Westbury, Wiltshire, went to London for a wedding and left her dog, Tina, with a neighbor, saying she would return by train at 10:30 P.M. In fact she returned five hours early, and was surprised to find tea waiting for her. Tina had suddenly jumped up and sat by the door wagging her tail. The neighbor, who knew that Tina often anticipated Sheila's return, rightly concluded that she had taken an earlier train.

Perhaps even more remarkable than dogs that know when their owners are coming home by train or bus are those that know in advance when their people will arrive by plane. There are many stories of this kind from World War II, when some pilots were allowed to keep their dogs at airfields. For example, Squadron Commander Max Aitken (later Lord Beaverbrook) kept his Labrador at the base of No. 68 Squadron. Edward Wolfe, who served under him, told me: "When the squadron was returning in ones and twos from an operation, his black Labrador, who would be sitting quietly in the mess, would get up and rush outside to meet his master. We always knew when Max Aitken was coming back."

I received a very similar report of a dog reacting to an owner who was a pilot in a glider squadron, where the returning planes were almost silent.

Anticipation by dogs belonging to airline staff are similarly impressive. A number of people who work for commercial airlines have found that their dogs know when they are coming home, even when no one else in the household knows. Elizabeth Bryan tells this story: "My whole working life has been as a cabin crew member working out of Gatwick Airport. For ten years my dog, Rusty, would jump around and bark at the same time I landed and then sit quietly watching the front door until I got home. The astonishing thing is there is no routine to my comings and goings, I could be gone one day or fourteen and no regular time of landing, yet he knew without fail."

Likewise, some people whose work takes them far away from home as passengers on planes have dogs that know when they are returning. Ian Fraser Ker, of Westcott, Surrey, first became aware of this phenomenon when he telephoned his wife on his arrival at Heathrow Airport. She told him she thought he would be coming because their dog, a Boxer, was very excited. "This developed so much that on days when my dog showed signs of excitement and would sit by the front door with his nose stuck as far into the letter box as it would go, my wife would actually cook lunch for me and, lo and behold, I would phone from the airport and say I was home."

In cases like these the dog could not possibly have recognized any familiar sounds or smells, or reacted to routines. And when people at home did not know when to expect the return, the dog could not have picked up its expectation from them. By a process of elimination, telepathy seems the most plausible explanation.

The alternative, as skeptics will hasten to point out, is that evidence based on pet owners' experiences cannot be trusted, because of tricks of memory, lying and deceit, or illusion and wishful thinking. Having talked to many pet owners about their experiences and interviewed members of their families, I have no reason to doubt that their accounts of the behavior of their dogs are generally trustworthy. And in the absence of any previous scientific investigations, these accounts are the only starting point we have if we want to explore this phenomenon.

It is right to maintain a skeptical attitude, ask further questions, and realize that people can be mistaken. But some people dismiss all the evidence from dog owners' experience as a matter of principle. This kind of compulsive skepticism stems from the dogma that telepathy is impossible. In my opinion such prejudices are barriers to open-minded scientific inquiry. They are not scientific but antiscientific. I am more interested in dogs than in dogma.

Obviously, it is necessary to follow up the study of case histories of dogs' anticipatory behavior with experimental investigations, as described later in this chapter. But it is also important to find out more about the natural history of dogs that know when their people are coming home. And since the evidence so far points to some kind of telepathic connection, we need to explore in more detail what the idea of telepathy might imply.

## Different patterns of telepathic response

"Telepathy" literally means "distant feeling," from two Greek roots: *tele,* as in "telephone" and "television," and *pathe,* as in "sympathy" and "empathy." If dogs are responding telepathically to their owners, they must somehow be picking up their owners' thoughts or feelings about going home. There are three main ways this might happen:

1. Some dogs might react only when their owners are nearing home and are of course aware of their own imminent arrival. Another way of expressing this might be to say that dogs feel their owners' approaching presence. A dog might react, say, two minutes or ten minutes before the owner's return, regardless of when the person set off.
2. Some people when traveling homeward may be thinking or feeling very little about going home for much of their journey; they may be fully engaged in conversation or some other activity. But there come stages in journeys at which feelings and thoughts turn homeward with increased intensity—when getting off a plane at an airport, for example, or disembarking from a ship or leaving a train or bus. Some dogs might pick up homeward-bound thoughts and feelings at this stage.
3. The most extreme manifestation of telepathy would occur if dogs were able to pick up their owners' *intention* to return, and if they reacted when the owners were setting off or even when they were getting ready to set off.

In fact all three types of anticipation are common. Some dogs anticipate their owners' return only a few minutes in advance. Perhaps the animal heard or smelled them, and telepathy might have nothing to do with it. But when a dog reacts more than five minutes in advance, the telepathic hypothesis needs to be taken seriously, especially if the dog still reacts when the windows are closed and its reactions do not depend on the wind direction, which would greatly influence the transmission of smells and sounds. And there are many cases where dogs regularly react ten minutes or more before a person comes home, irrespective of the wind direction. One example is Peter Edwards and his Irish Setters. Other examples are dogs at airports that react when their owner's plane is about to land, and dogs that meet their owners at bus stops, setting off when the bus is on the way.

Secondly, there are dogs that react when people get off boats, airplanes, trains, and buses and start the final part of their homeward journey. We have already seen examples of dogs that react when crew members and passengers on commercial flights arrive at the airport; and many others react when people get off boats, trains, or buses.

Finally, some dogs seem to react to people's intention to go home, even before they actually set off. Louise Gavit's dog, BJ, is one example. She had no regular schedule to her comings and goings. With the help of her husband observing BJ at home, she found that the dog typically reacted as follows:

> As I leave the place I have been, and walk to my car with the intent to come home, our dog, BJ, awakens from sleep, moves to the door, lies down on the floor near the door, and points his nose toward the door. There he waits. As I near the drive he becomes more alert and begins to pace and show excitement the nearer I move to home. He is always there to poke his nose through the crack, in greeting, as I open the door. This sensing seems to be unlimited by distance. He does not seem to respond at all to my leaving one place and moving to another, his response seems to become apparent at the time when I form the thought to return home, and take the action to walk toward my car to come home.

There is, of course, nothing new about this kind of behavior. It has been noticed and remarked upon for many years. In his well-known book *Kinship with All Life,* J. Allen Boone describes how the dog, Strongheart, anticipated his return from lunch at his club in Los Angeles some

12 miles away. A friend looked after Strongheart while Allen was out: "There was never any set time for my returning, but at the precise moment when I decided to leave the club and come home, Strongheart would always quit whatever he happened to be doing, take himself to his favorite spot for observation, and patiently wait there for me to turn the bend in the road and head up the hill."[4]

The same pattern of response has shown up in experiments. For example, Monika Sauer, who lives near Munich, Germany, carried out some tests at my request with her dog, Pluto, whose reactions were observed by her partner. Pluto reacted not only when Monika set off to come home in her own car but also when she set off in friends' cars with which he was unfamiliar. I then asked her to come home by taxi. When she did so, Pluto reacted forty minutes before her arrival. The journey took thirty minutes. She telephoned for the taxi and waited for it for ten minutes before setting off. The dog reacted not when she got into the taxi, but when she ordered it.

Advance reactions of this kind would go unnoticed unless people were paying close attention to the times at which they set off and the times at which the dogs reacted. Among those who do pay attention are Catherine and John O'Driscoll, whose Golden Retriever, Samson, is particularly sensitive to John's return. For example, one day John was at the theater in Northampton, England, when Samson dashed to the door excitedly, much sooner before his return than it would take for him to come home. Catherine told me: "I asked John what he was doing at the time, and he said he was looking at his watch wishing he could come home." On another occasion when John was at a meeting, "He was looking at his watch and closing his briefcase at the same time as Samson dashed to the door barking excitedly."

There are many other examples of this kind. In 218 (19 percent) of the 1,133 accounts on the database of dogs knowing when their owners are returning, they are said to react when the person sets off to come home or is preparing to do so.

Perhaps some of the dogs that appear to react only a few minutes before a person arrives home do in fact know when their owner sets off but show obvious signs of excitement only when the person is getting close. Earlier, more subtle responses may pass unnoticed.

## Returning from holidays and long absences

Most of the examples I have discussed so far concern dogs that respond when their owners are returning from work or from fairly brief absences. Now I turn to dogs' responses to their owners' return from a longer absence, such as a vacation. Some dogs do not anticipate their owners' arrival when they have just gone out for the day, but they react when the person has been away for longer periods. Take, for example the Marchioness of Salisbury's dog, Jessie, who lived with her at Hatfield House in Hertfordshire (Figure 2.1).

"Jessie is a very acute and intelligent little dog," Lady Salisbury said. "She always seems to know what I'm going to do almost before I do myself." When Lady Salisbury went abroad, she usually left Jessie, a hunt terrier, with her head gardener, David Beaumont. He and his wife generally knew when Lady Salisbury was on her way home, because Jessie became restless and waited by the door or by the gate of their house for hours before she arrived. Jessie's behavior was documented by Miriam Rothschild, the distinguished naturalist, who kindly passed on her observations to me. On one occasion, for example, Jessie began to react when Lady Salisbury was packing and preparing to leave a house in Ireland; on another, when she was leaving for the airport in Cracow, Poland. Lady Salisbury says that the dog's mother was even more sensitive to her homecomings than Jessie was; she responded even if the marchioness was away for only a day. Jessie did not react unless she had been away for at least three days.

Sometimes the behavior of the dog seems to be related to the owner's thoughts and intentions well before the person actually begins the journey. This was the case with Frank Harrison, who was taken ill with a fever soon after he joined the British Army and, on his discharge from the hospital, was given a few days' sick leave. He did not inform his parents.

> When I arrived home Sandy, our Irish Terrier, was by the door and I was told that he had not moved from the door for two days, except to be fed and exercised. This was about the time I had been told I was

***Figure 2.1** The Marchioness of Salisbury with her hunt terrier, Jessie, at Hatfield House, Hertfordshire. (Photograph: Phil Starling)*

being given sick leave. His behavior had naturally caused concern to my parents. When I unexpectedly arrived home my mum said, "He knew you were coming. That explains it." This waiting at the door happened throughout my two and a half years of service in the army. Sandy would move to the door about forty-eight hours before I came home. My parents knew I was coming because Sandy knew.

I have received more than twenty other accounts of dogs anticipating the arrival of young men coming home on leave from the armed forces or the merchant navy, and in many cases the families were not informed in advance. Sometimes the dogs reacted one or two days before the young man arrived home, as Sandy did, and sometimes a few hours.

The anticipatory reactions of dogs when their owners set off to come home from another continent suggest that telepathic communication can occur over great distances. It does not seem to fall off with distance in the way that gravitational, electrical, and magnetic phenomena do.

In some instances, a dog's anticipation can be pinpointed to a particular stage of preparing or setting off. Tony Harvey was returning to his farm in Suffolk from a three-week shooting holiday on Dartmoor, 250 miles away, having left Badger, his Border Terrier, with his wife at home. When he returned, his wife told him that Badger had jumped out of his basket and onto the windowsill at 6:40 A.M. "This was exactly the time that I had started home from Dartmoor," Harvey said. "This was not the time we started to load up, but the exact time that the lorry started along the road for home." Badger was "excited all day, standing on the window and looking up the yard," until his owner finally arrived at 9:30 P.M.

As in the case of people returning from work, some dogs react when the person is nearing home, rather than at the beginning of the owner's journey. For example, when Her Majesty Queen Elizabeth II visits her estate at Sandringham, England, the staff does not need to be told when she is approaching because her gundogs have already alerted them. "All the dogs in the kennels start barking the moment she reaches the gate—and that is half a mile away," said Bill Meldrum, the head gamekeeper. The Queen is famous for her affinity to her animals, and training her gundogs is one her favorite pursuits.[5]

The people who work in kennels have frequent opportunities to observe the behavior of dogs prior to the return of their owners from vacations or journeys. My associates and I have interviewed kennel proprietors in both Britain and the United States and have found that most have noticed that some dogs seem to know when they are about to go home. Typical comments included the following: "Some get more alert when it is the day for them to go home." "There's an air of expectancy a few hours beforehand." "Some of the dogs do act different on the day when they are going home." One kennel proprietor in eastern England, however, firmly denied that anything like this occurred at her establishment. "Dogs are so happy here that they soon forget all about their owners and have no interest in their return." This was, however, an isolated observation.

Perhaps some dogs in kennels behave differently because the kennel workers give them more attention when they are about to be picked up. But sometimes owners return unexpectedly early, and some dogs still seem to know. Sam Hyer, of Rockford, Michigan, recalled that one "relatively calm dog . . . stood by the door, for three hours. I took it out several times, but it did not need to relieve itself and in two hours his owner drove in, two days early. I had no idea the owners would be early."

## The bonds between dog and person

Most people whose dogs anticipate their arrival feel that they have a close bond, they enjoy a strong emotional connection, or they are very attached to the dog. In 79 percent of the cases on our database, the dogs respond only to one person, 14 percent respond to two people, and only 7 percent react to three or more. When dogs respond to more than one person, these are usually members of the family. Almost the only other people whose arrival dogs anticipate are friends whom they are particularly fond of, or people who take them for walks or bring them treats.

The only exceptions are cases in which the dog has a strong aversion to a particular person. John Ashton, for example, had a friend who disliked dogs who used to visit him at his home in Lancashire about once a week. At first his German Shepherd, Rolf, usually a good-natured dog, showed no unusual behavior. "After a few months, my friend Clive visited me one night and about ten minutes before he arrived, Rolf was at my garage waiting and growling and had to be restrained when Clive got there. I can only assume that Clive had smacked or kicked him away on his previous visit. After this night, I always had to go to the garage and meet Clive and restrain Rolf. He always knew ten to fifteen minutes before Clive's time of arrival."

In one very interesting case a Springer Spaniel reacted differently depending on the intention of the visitor. Christopher Day, a vet in Oxfordshire, was the visitor, and the dog belonged to his mother-in-law:

> The dog used to know whether I was visiting socially or whether I was visiting as a vet. She would be all over me and whooping with delight if I was visiting socially, but if I visited as a vet she was hiding behind the boiler. There was nothing I could see which would give her a clue that I was visiting as a vet, and anyway she would have made the decision to hide before I came into the house. She got it right every time. I used to visit quite often, pop in and do all sorts of things, although as a vet I'd visit very rarely. And I didn't just visit as a vet because the dog was ill, sometimes it could be routine things. But the dog knew when I was on duty and when I wasn't.

Thus the ability of dogs to know when people are coming seems to depend on an emotional bond, usually positive but sometimes negative, and the dog's behavior can be influenced by the person's reason for coming. But generally speaking, the reaction depends on affectionate relationships with the dog's human companions and with visiting family members and close friends.

It is of course well known that dogs can form strong bonds with people. James Serpell, who pioneered the study of human-dog relationships at Cambridge University, expressed it as follows: "The average dog behaves as if literally 'attached' to its owner by an invisible cord. Given the opportunity, it will follow him everywhere, sit or lie down beside him, and exhibit clear signs of distress if the owner goes out and leaves it behind, or shuts it out of the room unexpectedly."[6]

I think that the evidence considered in this and the following chapters suggests that the invisible cord connecting dog to owner is elastic: it can stretch and contract (see Chapter 1, Figure 1.5B). It connects dog and owner together when they are physically close to each other, and it continues to attach dog to owner even when they are hundreds of miles apart. Through this elastic connection, telepathic communication takes place.

## Telepathy or precognition?

Many pet owners whose animals know when a member of the family is coming home ascribe the dog's behavior to telepathy, to a sixth sense, or to extrasensory perception.

The term "telepathy" implies that the dog is reacting to the thoughts, feelings, emotions, or intentions of a distant person. But the terms "sixth sense" and "ESP" are more general. They may be used to mean "telepathy," but they are also often used in connection with a variety of other unexplained phenomena, including the ability to anticipate danger and the ability to find the way home. And some of the phenomena ascribed to the sixth sense or ESP seem to include precognition—that is, knowing beforehand about future events.

Could it be that dogs know when their owners are coming home because of a precognition of the actual arrival, rather than because they are picking up the thoughts or intentions of their owners?

Perhaps in some cases this is so. But telepathy seems to me a more likely explanation when dogs respond at the time their owners are setting off or are simply intending to set off. Telepathy also seems a more likely explanation for the responses of animals when their people reach a crucial stage in their journey home, such as disembarking from a plane, boat, train, or bus.

## What happens when people change their minds?

One way of distinguishing the possible roles of telepathy and precognition is to look at what happens when people change their minds. If they set off homeward and then their journey is interrupted, what happens? If the animal's response is precognitive—that is, if the pet foresees the person's arrival—it should not react when a person's journey is aborted. If it is responding telepathically, it should react to the intention to come home, even if the person does not arrive. So what actually happens?

One of the first examples I came across of a dog's reaction when someone changed his mind was told to me by Radboud Spruit, of Utrecht University, Holland. He was living about a six-minute drive away from his parents and used to visit them several times a week at irregular times. His mother noticed that the dog usually started to wait for him at the garden gate a few minutes before he actually began his journey. "One day my mother called me and asked if I had planned to visit them the day before, because the dog waited for me. I had planned to visit them, but I changed my mind on the way. It was at the same time that our dog was waiting for me. My mother told me the dog got confused after fifteen minutes when I didn't arrive. It ran into the house and after some minutes it ran again to the gate. After about half an hour it looked as if the dog had forgotten about it."

In some cases, the people can tell precisely when a dog's owner set off and when she changed her mind. For example, Michael Joyce looked after his sister-in-law's dog while she and his wife went shopping in Colchester, Essex, 14 miles away. He noticed that at 4:45 the dog walked to the window and sat there. "Just a few minutes later it resumed its former position, sprawled out on the carpet. Then at about 5:15, half an hour later, it became excited and anxious again and remained near the window, waiting/anticipating their arrival. When my wife and sister-in-law arrived I said, 'You decided to leave Colchester at about 4:45, changed your minds, and later decided to leave at 5:15!' " This was indeed what had happened.

The evidence from these and other interrupted journeys supports the idea that dogs respond to their owners telepathically rather than by precognition of their arrival.

## How common are dogs that know when their owners are returning?

The people who have written to me in response to appeals for information tend to be those whose animals behave in particularly impressive ways. Obviously, people whose animals do *not* respond do not write in to say so. Hence my database does not contain a representative sample of all dogs and does not in itself reveal how common this anticipatory behavior is.

From informal surveys of friends, colleagues, and people who attend my lectures and seminars I found that between one-third and two-thirds of dog owners said they had noticed anticipatory behavior in their dogs. Readers can easily carry out their own surveys and see if they come up with similar results.

Although such informal surveys give a rough indication, they are open to a number of criticisms, the most important of which is that the people who are asked represent a biased sample. In order to avoid such possible bias, it is necessary to question a random sample of the population, using standard surveying techniques. My associates and I have completed four such surveys, which we carried out in very different geographical and cultural environments: in North London; in Ramsbottom, a town near Manchester in northwest England; in Santa Cruz, a beach and university town in California; and in the suburbs of Los Angeles in the San Fernando Valley.

We surveyed a random sample of households by telephone. The percentage with dogs in Santa Cruz and Los Angeles was 35 percent, close to the U.S. national average. In Ramsbottom it was 31 percent, slightly above the British national average. In London it was only 16 percent, in agreement with the tendency for dog ownership to be lowest in large cities, where more people live in apartments.

The first question pet owners were asked was "Have you or has anyone in your household ever noticed your animal getting agitated before a family member arrives home?" Those who answered yes were then asked: "How long before the person arrives is the pet agitated?" (They were then asked further questions about their pets, which I will discuss in Chapters 7 and 8.) Readers interested in the details of these surveys can read more about them in our papers in scientific journals[7] and on my website.[8]

In spite of the great differences between the places surveyed and the fact that the surveys were carried out by different people, the results are in remarkable agreement (see Figure 3.1, page 73). About half the dogs were said to show anticipatory behavior before their people came home; the overall average was 51 percent. The highest percentage was in Los Angeles (61 percent), and the lowest was in Santa Cruz (45 percent). These figures may have underestimated the positive responses, because people who live alone do not usually know whether or not their animal anticipates their return.

Most dogs that anticipated their owners' return did so less than ten minutes before the person arrived, but between 16 and 25 percent were said to do so more than ten minutes in advance.[9] Such reactions are unlikely to be due to sounds and smells, as I have discussed earlier, although some might be explicable in terms of routine.

No formal random surveys have yet been conducted in any other countries, but my own informal surveys in Belgium, Brazil, Canada,

Denmark, France, Germany, Holland, Ireland, Norway, Portugal, and Switzerland have yielded results similar to those resulting from informal surveys in Britain and America.

## Why do so many dogs *not* react?

Even if, as my research indicates, about half the dogs in a given place anticipate their owners' arrival, there are still about half that do not. Why not? I can think of five possible explanations:

First, when people live alone there is no one to observe the dog's reactions, so the reactions would pass unnoticed.

Second, some dogs may have reacted in the past, but their owners failed to notice or offer any encouragement. In households where people *do* notice this behavior, simply paying attention to it may encourage the dog. But in many households there is no incentive for the dog to show what it knows. If more owners paid attention to this behavior, the percentage of dogs showing it might rise.

Third, the bond between the dog and its owner may not be strong enough to evoke this behavior. The dog may not be sufficiently interested in the person's return.

Fourth, some dogs may be less sensitive than others. There is a wide variation of sensitivity in all other respects, including smell, hearing, and sight, even among dogs that are closely related. So why not in this?

Fifth, some breeds may be relatively insensitive.

These possibilities are mutually compatible, and all could work together.

Too little is known at present to test the first four possibilities. But the fifth can be explored straightaway. There is already enough information in the database and from the formal surveys to investigate whether some breeds are more sensitive than others.

## Are some breeds more sensitive than others?

I have received reports of anticipatory behavior by 102 distinct breeds of dogs as well as many crossbreeds and mixed breeds of unknown ancestry.

Dog breeds are commonly grouped into several groups, and different experts use different systems, which are more or less arbitrary. I use the following categories:

- **Sporting dogs.** This group includes pointers, retrievers, spaniels, and setters.
- **Hounds.** The two subgroups are sight hounds, such as Greyhounds and Afghan Hounds, and scent hounds, such as Bloodhounds and Foxhounds.
- **Terriers.**
- **Working and herding dogs.** This group consists of guard dogs, sled dogs, and dogs that once worked with livestock, such as Collies, German Shepherds, and Shetland Sheepdogs.
- **Nonsporting dogs.** This miscellaneous group includes Poodles, Dalmatians, and Bulldogs.
- **Toy dogs.** These small house dogs that have served as companion animals include Pekingese, Cavalier King Charles Spaniels, and Chihuahuas.

Out of the 657 accounts of anticipatory behavior on the database where the breed of the dog was given, the breakdown according to these categories is as follows:

| | |
|---|---|
| Sporting dogs | 129 |
| Hounds | 71 |
| Terriers | 113 |
| Working and herding dogs | 207 |
| Nonsporting dogs | 57 |
| Toy dogs | 32 |
| Mixed-breed dogs | 48 |

The individual breeds that occur most frequently in these reports are German Shepherds (73 examples), Labrador Retrievers (57), Collies (57), Poodles (48), and Dachshunds (41). But this may not mean they are unusually sensitive; it may simply reflect the fact that these are some of the most popular breeds. Likewise, the fact that most reports concern working dogs and sporting dogs may simply reflect the fact that more people keep dogs from these groups than from the other categories.

So although no detailed conclusions can be drawn from the reports on the database about the sensitivity of different kinds of dogs, it is clear that anticipatory behavior is widespread and is not confined to any particular group.

The formal surveys carried out in Britain and the United States give a more reliable picture, because they are based on random samples. The combined results from all four surveys are shown in the table below. In addition to the totals for each group, figures for particular breeds are given in cases where there were more than ten dogs of that breed:

| KIND OF DOG | TOTAL NUMBER SURVEYED | NUMBER ANTICIPATING ARRIVALS | PERCENTAGE ANTICIPATING ARRIVALS |
|---|---|---|---|
| **Sporting dogs** | 58 | 30 | 52 |
| Labrador Retrievers | 21 | 8 | 38 |
| Spaniels | 21 | 12 | 57 |
| **Hounds** | 12 | 6 | 50 |
| **Terriers** | 41 | 23 | 56 |
| **Working and herding dogs** | 55 | 24 | 44 |
| German Shepherds | 16 | 6 | 38 |
| Collies | 13 | 8 | 62 |
| **Nonsporting dogs** | 17 | 11 | 65 |
| **Toy dogs** | 20 | 13 | 65 |
| **Mixed-breed dogs** | 82 | 39 | 48 |

None of the differences between groups are statistically significant, and they could simply be chance variations owing to the relatively small size of the sample. Therefore not much can be concluded from these differences, but I suspect that the relatively high percentages in the toy and nonsporting groups might be repeated in other surveys. Many of the breeds in these groups have been bred over generations to serve as companions for their owners. They may tend to be more sensitive to their owners' intentions because of their breeding and because they are more likely to be kept indoors. They may literally be closer to their owners than large dogs, a higher proportion of which are kept in kennels outdoors or restricted to certain parts of the house.

These figures confirm that many kinds of dogs seem to anticipate the arrival of their owners. This ability is not confined to any particular breed or group.

Nor is it confined to one sex, although males tend to show this behavior more than females. Of the 618 accounts on the database of dogs where the sex of the dog is mentioned, 54 percent were males. In the random household surveys in Britain, 52 percent of the dogs said to show anticipatory behavior were males.

## Logs of dogs' behavior

Dog owners' reports of their pets' behavior are an invaluable starting point for further investigation. In fact they are the only possible starting point, since in the absence of any scientific investigations, they are the only information available.

The next step is the keeping of written records of dogs' behavior. Much can be learned from such logs, and the only equipment needed is a notebook and pen. But for more rigorous research, it is essential to conduct controlled experiments and to film the dogs' responses on time-coded videotape. Such investigations are the subject of the rest of this chapter.

At my request, more than twenty owners kept logs of their dogs' behavior prior to the return of a member of the family, and some carried out experiments by coming home at unusual times and traveling in an unfamiliar vehicle.

These logs are extremely illuminating and reveal details about the animals' behavior that would otherwise have been forgotten. They confirm that some dogs do indeed anticipate people's arrival fairly reliably, though not necessarily on every occasion. I would encourage readers whose animals seem to anticipate their arrival to keep logs themselves, noting down

1. The date and exact times at which the animal seems to show anticipatory reactions.
2. The time at which the person returns and the time at which he or she sets off to come home.

3. Where the person went and how long he or she was away.
4. How the person traveled home.
5. Whether or not the person arrived at a routine or expected time.
6. Any comments or observations.

These records are best kept in a special notebook. It is important to note the animal's failures as well as successes, so that if the dog shows no signs of anticipation before the person arrives home, this is duly recorded. False alarms should also be noted, that is, occasions when the dog seemed to be anticipating an arrival when the owner was not coming home.

In all but one of the logs I received, the dogs were regularly reacting ten minutes or more in advance of the person's arrival; some reacted hours in advance, when their person was setting off on a long homeward journey. These reactions cannot be explained in terms of hearing or smelling the returning person. Most cannot be accounted for in terms of routine, either. However, in one of the logs, the dog usually anticipated its owner's arrival by only three or four minutes, so it is just possible that on these occasions it could have heard her car approaching.

In several cases dogs appeared to give false alarms, but then it turned out that their people had indeed set off to come home and had changed their minds or been interrupted on the way.

Sometimes the dogs failed to react in advance of their owners' return when they were distracted, sick, or frightened. Sometimes they failed to react for no apparent reason. But on the great majority of occasions, the dogs anticipated their people's arrival by ten minutes or more.

The most extensive set of records concern a male mixed-breed terrier called Jaytee, who lived in northwest England with his owner Pamela Smart (Figure 2.2).

## Jaytee's anticipations

Over several years, Jaytee was observed by members of Pamela Smart's family to anticipate her arrival by half an hour or more. He seemed to know when Pam was on her way even when no one else knew and even when she returned unexpectedly.

*Figure 2.2 Pam Smart with Jaytee. (Photograph: Gary Taylor)*

Pam adopted Jaytee from the Manchester Dogs' Home in 1989 when he was still a puppy and soon formed a close bond with him. She lived in Ramsbottom, Greater Manchester, in a ground-floor flat, next door to her parents, William and Muriel Smart, who were retired. When she went out, she usually left Jaytee with her parents.

In 1991, when Pam was working as a secretary in Manchester, her parents noticed that Jaytee used to go to the French window almost every weekday at about 4:30 P.M., around the time she set off to come home. Her journey usually took forty-five to sixty minutes, and Jaytee would wait at the window most of the time that she was on her way. Since she worked routine office hours, the family assumed that Jaytee's behavior depended on some kind of time sense.

Pam was laid off in 1993 and was subsequently unemployed. She was often away from home for hours at a time, and was no longer tied to any regular pattern of activity. Her parents did not usually know when she would return, but Jaytee continued to anticipate her return. His reactions seemed to occur around the time she set off on her homeward journey.

In April 1994, Pam read an article in the *Sunday Telegraph* about the research I was doing on this phenomenon[10] and volunteered to take part. The first stage in this investigation was the keeping of a log by Pam and her parents. Between May 1994 and February 1995 on 100 occasions she left Jaytee with her parents when she went out, and they made notes on Jaytee's reactions. Pam herself kept a record of where she went, how far she traveled, what mode of transport she used, and what time she set off to come home. On 85 of these 100 occasions, Jaytee reacted by going to wait at the French window before Pam returned, usually ten or more minutes in advance.

When these data were analyzed statistically, they showed that Jaytee's reactions were very significantly[11] related to the time that Pam set off, as if he knew when she was starting to go home.[12] It did not seem to matter how far away she was.[13]

However, Jaytee did not react on 15 out of 100 occasions. Was there anything unusual about these occasions? On some, Pam's mother was away from home or asleep. Jaytee was closely bonded with Mrs. Smart but rather afraid of Mr. Smart. When left alone with him, Jaytee hid in the bedroom and was not observable. On some occasions, there were major distractions, such as a bitch in heat in a neighbor's flat. On some occasions Jaytee was sick. But on three occasions there were no apparent distractions or reasons for his lack of response. Thus Jaytee did not always react to Pam's return, and he could be distracted.

Jaytee's anticipatory reactions usually began when Pam was more than 4 miles away, and in some cases more than 40 miles away. He could not possibly have heard her car at such distances, especially when the car was downwind and against the background of the heavy traffic in Greater Manchester and on the M66 motorway, which runs close to Ramsbottom. Moreover, Mr. and Mrs. Smart had already noticed that Jaytee anticipated Pam's return even when she arrived in unfamiliar vehicles.

Nevertheless, to check that Jaytee was not reacting to the sound of Pam's car or other familiar vehicles, we investigated whether he still responded when she traveled by unusual means: by bicycle, by train, and by taxi. He did.[14]

Pam did not usually tell her parents in advance when she would be coming home, nor did she telephone to inform them. Indeed, she often did not know in advance when she would be returning after spending an evening out, visiting friends, or shopping. But her parents might in some cases have guessed when she would be coming and then, consciously or unconsciously, communicated their expectation to Jaytee. Some of his reactions might therefore be due to her parents' anticipation rather than to some mysterious influence from Pam herself.

To test this possibility, we carried out experiments in which Pam set off at times selected at random after she had left home. These times were unknown to anyone else. In these experiments, Jaytee started to wait when she set off, or rather a minute or two before while she was making her way to her car, even though no one at home knew when she would be coming.[15] Therefore his reactions could not be explained in terms of her parents' expectations.

By this stage it was clearly important to start taping Jaytee's behavior so that a more precise and objective record could be kept. And just at this point I was approached by the Science Unit of Austrian State Television (ORF) who wanted to film an experiment with a dog. Pam and her parents kindly agreed to do this filmed experiment with Jaytee.

Together with Dr. Heinz Leger and Barbara von Melle of ORF, I designed an experiment using two cameras, one filming Jaytee continuously in Pam's parents' house, and the other following Pam as she went out and about.

This experiment duly took place in November 1994. Neither Pam nor her parents knew the randomly selected time at which she would be asked to return.

Some three hours and fifty minutes after she had set out, she was told it was time to go home. She then walked to a taxi stand, arriving there five minutes later, and reached home ten minutes after that. As usual, Jaytee greeted her enthusiastically.

From the videotapes, Jaytee's behavior can be observed in a detail not previously possible. During the period that Pam was out, he spent practically all the time lying quite calmly by the feet of Mrs. Smart. In the edited version produced by ORF for transmission on television, over the period that Pam was told to return, both videotapes can be seen together on a split screen in exact synchrony, so that Pam can be observed on one side of the screen, and Jaytee on the other. To start with, Jaytee is, as usual, lying by Mrs. Smart's feet. Pam is then told that it is time to return, and almost immediately Jaytee shows signs of alertness, with his ears pricked. Eleven seconds after Pam has been told to go home, while she is walking toward the taxi stand, Jaytee gets up, walks to the window, and sits there expectantly. He remains at the window for the entire duration of Pam's return journey.

There seems no possible way in which Jaytee could have known by normal sensory means at what instant Pam was setting off to come home. Nor could it have been routine, since the time was chosen at random and was at a time of day when Pam would not normally have returned.

This experiment highlights the importance of Pam's intentions. Jaytee started to wait when Pam first knew she was going home, before she got into the vehicle and began the taxi journey. Jaytee seemed to be responding telepathically.

## Videotaped experiments with Jaytee

In April 1995, I received a grant from the Lifebridge Foundation, of New York, to support my research on the unexplained powers of animals. By then, as a result of the publication of my book *Seven Experiments That Could Change the World*[16] and appeals for information from pet owners, I was receiving hundreds of letters. I read and acknowledged them all personally, but I was unable to cope on my own with the task of organizing them on a database. I needed a research assistant who had the necessary secretarial and computer skills to build up the database, who was interested in animals, and who was capable of carrying out experiments on her own. Pam Smart fitted the job description ideally.

So after a year's voluntary research with her own dog, Pam became my full-time research assistant. The experiments with Jaytee continued, but now with the regular videotaping of Jaytee's behavior throughout the whole period that Pam was out.

The procedure was kept as simple as possible, so that observations of Jaytee could be done routinely and automatically. The video camera was set up on a tripod, and left running continuously in the long-play mode with a long-play film, with the time code recorded on it. The camera was pointing at the area where Jaytee usually waited, by the French window in her parents' flat. These experiments were possible only because her parents kindly agreed to have their living room continuously monitored for hours on end, sometimes several times a week. They and the members of their extended family who often visited them simply got used to it and carried on life as usual.

Jaytee's behavior was also videotaped in Pam's own flat while he was on his own and in the house of her sister Cathie. The videotapes were scored by a third person who did not know any details of the experiment. In most videos, for most of the time, Jaytee was not on camera. But every time he appeared by the window, the exact time he did so was recorded, as was the length of time he stayed there. Notes were also made about his behavior. For example, on some visits to the window, he was obviously barking at passing cats, or watching other activities outside. On others he was sleeping in the sun. On others he looked as if he was just waiting. We made and analyzed over 120 videotaped records of Jaytee's behavior from the time Pam left home until the time she returned.

Between May 1995 and July 1996 we made thirty videotapes of Jaytee at Pam's parents' flat while Pam went out and about. Pam's parents were not told when she would be returning, and she usually did not know exactly herself. The purpose was to observe how Jaytee behaved under more or less natural conditions. Seven of these thirty videotapes were taken in the daytime, at various times in the morning and afternoon. Twenty-three were taken in the evening, with Pam returning at different times between 7:30 and 11:00.

The overall results are shown in Figure 2.3A. The general pattern is clear. On average, Jaytee waited at the window much more when Pam was on her way home, and he began waiting while she was preparing to set off. He was at the window much less during the main period of her absence. These differences were highly significant statistically ($p < 0.000001$)[17] and show that Jaytee was reacting to Pam's intentions.

Jaytee's pattern of response can be seen in more detail in the graphs in Figure 2.4. However long Pam's absence, Jaytee waited by the window much more when she was on the way home than at any other period. He usually began to wait there shortly before she set off, while she was thinking about going home and preparing to do so.

At an early stage in our research, we found that Jaytee anticipated Pam's return even when she set off at randomly selected times, but this was such an important finding that we carried out a further series of twelve videotaped experiments in which Pam returned at random times. I selected a time at random by throwing dice, and when this time came I beeped her through a telephone pager. She then set off as soon as possible. As usual, a videotape was made of the area by the window throughout her entire absence.

The results, summarized in Figure 2.3B, show the same general pattern as Pam's ordinary homecomings (Figure 2.3A), and confirm that Jaytee's reactions were not a matter of routine or of expectations communicated by her parents. Jaytee was at the window far more when Pam was on her way home than during the main part of her absence (55 percent of the time as opposed to 5 percent).

Most of the experiments with Jaytee were in the home of Pam's parents, but we also carried out additional videotaped experiments with Jaytee left on his own in Pam's flat. The overall pattern was similar: Jaytee waited at the window significantly more when Pam was on the way home than when she was not. But the percentage of time Jaytee spent at the window was lower than in Pam's parents' flat.

In Pam's sister's house, in order to look out of the window Jaytee had to balance on the back of a sofa. He could not wait there comfortably and rarely stayed for long. Nevertheless, in a series of videotaped experiments, the general pattern of his response was similar to that in Pam's parents' flat, although again the percentage of time he spent at the window was lower.

We also made a series of videotapes on evenings when Pam was not coming home until very late at night or not at all. These served as controls and showed that Jaytee went to the window very little at any stage during the four-hour control period.

Following the successful experiment with Jaytee carried out by ORF, a number of reports about this research appeared on television and in newspapers. Journalists sought out a skeptic to comment on these results, and several chose Dr. Richard Wiseman, who regularly appears on British television as a debunker of psychic phenomena. He is a psychologist at the University of Hertfordshire and a consulting editor of the *Skeptical Inquirer,* the organ of the Committee for the Scientific Investigation of Claims of the Paranormal. Wiseman suggested that Pam

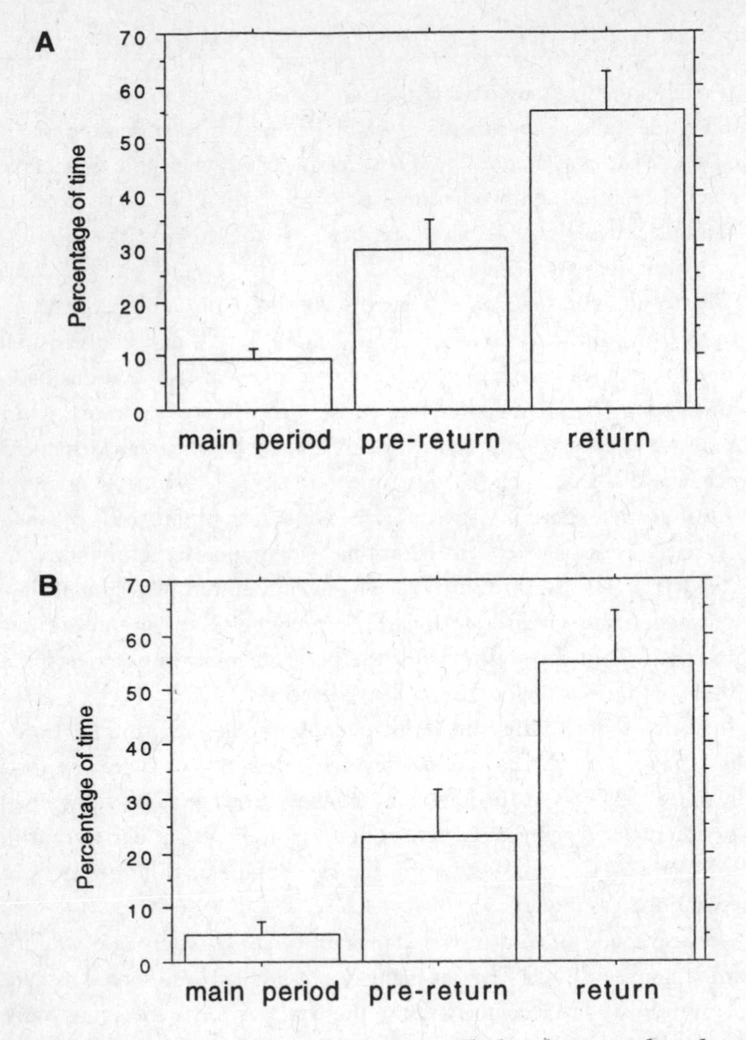


*Figure 2.3 Jaytee's reactions to Pam's returns. The bar diagrams show the percentage of time that Jaytee spent by the window during the main part of Pam's absence ("main period"), during the ten minutes prior to her setting off to come home ("pre-return"), and during the first ten minutes of her journey home ("return"). (The standard error of each value is indicated by the bar at the top.)*

*A: Averages from thirty experiments in which Pam returned home at times of her own choice.*

*B: Averages from twelve experiments in which Pam returned home at randomly selected times in response to being beeped on her pager.*

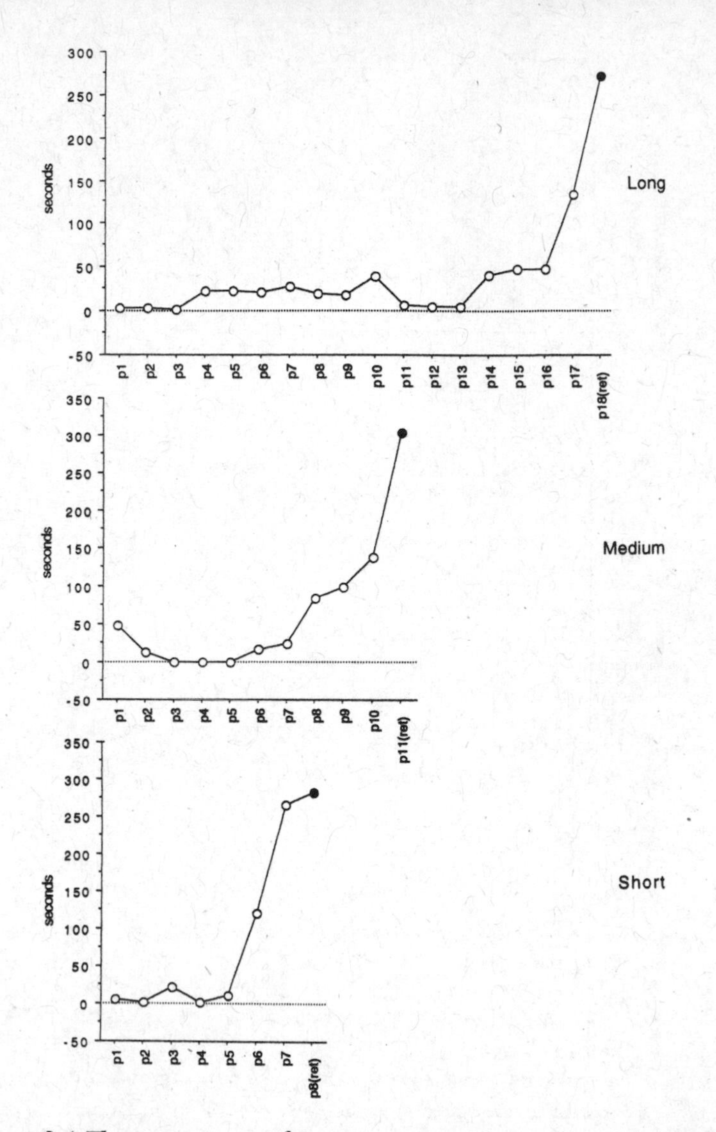


*Figure 2.4 The time courses of Jaytee's visits to the window during Pam's long, medium, and short absences. The horizontal axis shows the series of 10-minute periods (p1, p2, etc.) from the time she went out until she was on her way home. The last period shown on the graph represents the first 10 minutes of Pam's return journey ("ret"), the point for which is indicated by a filled circle (●). The vertical axis shows the average number of seconds that Jaytee spent at the window during each 10-minute period. The graphs represent the average of eleven long, seven medium, and six short experiments.*

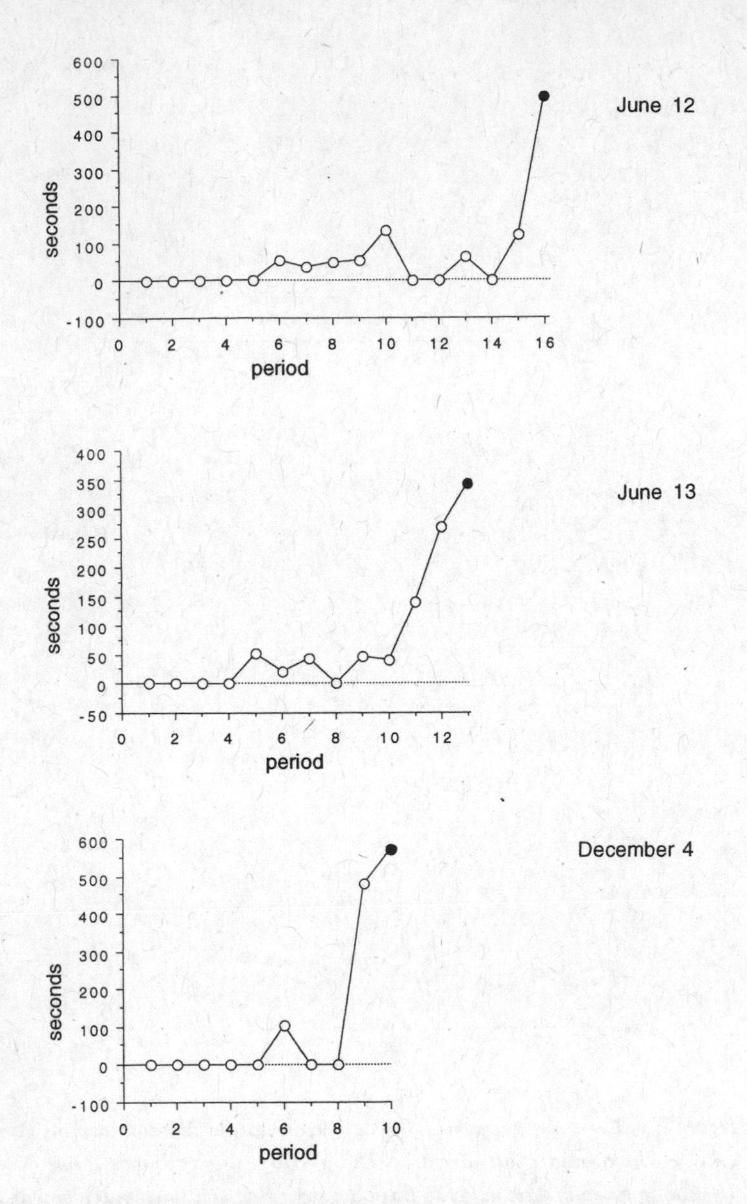


*Figure 2.5 Results of the three experiments carried out by Richard Wiseman and Matthew Smith with Jaytee at Pam's parents' flat in 1995. The graphs show the amount of time that Jaytee spent at the window in successive 10-minute periods. As in Figure 2.4, the final point on each graph represents the first 10 minutes of Pam's return journey, and is indicated by a filled circle (●). (Graphs plotted from the data of Wiseman, Smith, and Milton, 1998.)*

returned at routine times, or that Jaytee picked up from her parents when she would return, or that Jaytee recognized the sound of her car at a distance. We had already tested and eliminated these possibilities, but our evidence did not convince Wiseman. So I invited him to do some tests of his own, and Pam and her family kindly agreed to help him.

In his experiments, Wiseman himself videotaped Jaytee while his assistant, Matthew Smith, went out with Pam and videotaped her. They traveled either in Smith's car or by taxi, and went to pubs or other places 5 to 11 miles away. Smith then selected a number at random to determine when they would set off home, or else telephoned a third party who had selected a number at random for him. Thus he himself knew in advance when they would set off, but he did not tell Pam until it was time to go.

The videotapes of Jaytee's behavior were analyzed blind by someone who did not know when Pam had set off to come home. Pam and I have also analyzed them. We all agree about when Jaytee went to the window and how long he stayed there.[18] These results are shown in Figure 2.5. The pattern was very similar to that in my own experiments, and confirmed that Jaytee anticipated Pam's arrival even when she was returning at a randomly chosen time in an unfamiliar vehicle. However, this is not the conclusion that Wiseman and Smith drew. They announced to the world through press releases that they had *refuted* Jaytee's abilities![19] It was only in 2009 that Wiseman finally conceded that his results showed the same pattern as my own.[20] I describe my controversy with Richard Wiseman in the Appendix, together with an account of other interactions with skeptics.

## Experiments with Kane

Following the experiments with Jaytee, Pam and I ran a series of tests with another return-anticipating dog, Kane, an eighteen-month-old male Rhodesian Ridgeback (Figure 2.6). He lived in Middleton, a town in Greater Manchester, with his owner, Sarah Hamlett, and her partner, Jason Hopwood. Several months prior to the study, Jason noticed that Kane seemed to know when Sarah was coming home. The dog would look out of a window when Sarah was on her way, standing on his hind legs with his front paws resting on a table in front of the window (Figure 2.6B).

*Figure 2.6 Pictures from the videotape of a trial with Kane on July 29, 2008, showing Kane lying on the sofa five minutes before his owner set off to come home (A), and then Kane at the window five minutes after his owner set off homeward, eighteen minutes before she arrived (B).*

The window overlooked the road on which Sarah approached their ground-floor apartment, but the road was partially obscured by a hedge, and approaching cars were visible only when they were less than 300 feet away.

We carried out a series of ten trials in which Sarah drove at least 5 miles away by car. During her absence the area by the window was filmed continuously on time-coded videotape. Sarah came home at nonroutine times. As a student, she had to attend college at different times of the day, and she also went to visit her horse, worked in her father's shop, and volunteered at several veterinary clinics. She did not tell anyone when she would return; indeed she did not usually know in advance herself. In some of the trials she set off at times Pam randomly selected after the experiment began. Pam told Sarah by pager when it was time to go home.

In nine out of ten trials Kane spent the most time at the window when Sarah was on her way home. On average he was at the window 26 percent of the time while she was returning and only 1 percent of the time throughout the rest of her absence. This difference is highly significant statistically.[21]

These formal tests with Kane and Jaytee validate what many people have already observed, and the results confirm that some dogs anticipate the return of their owners at nonroutine times, in unfamiliar vehicles, when no one at home knows when the person will return, and when the owner is miles away, beyond the range of the senses of sight, hearing, and smell. Many cats act similarly.

*Chapter 3*

# Cats

Many cats lead a double life: outdoors they are solitary hunters; indoors they are more or less affectionate companions. In relation to their human keepers, they behave rather like kittens do toward the mother cat that feeds and protects them.

In general, cats are obviously more independent and less sociable than dogs. A cat usually feels no need to be near its owner all the time. While most dogs are person-centered, most cats are home-centered.

Cats have lived in close association with human beings for at least 9,000 years. They were probably first domesticated in North Africa, and their wild ancestor was the African wildcat *Felis silvestris*, subspecies *libyca*. The archaeological evidence suggests domestication began in the same places and time as the development of year-round human settlements, with the beginning of an agricultural economy. Species such as mice soon started living in human village environments, attracted by garbage and grain stores, and it seems likely that native wildcats also adapted to living in human settlements because of the plentiful supply of mice.[1] Some of these semi-wildcats became pets, and certainly by 3,600 years ago the ancient Egyptians revered them and kept them in their houses. They were depicted in tomb paintings and were believed to embody the cat-goddess Bastet, related to the terrifying lioness goddess Sekhmet.

Rudyard Kipling's famous Just So story *The Cat That Walked by Himself* epitomizes feline characteristics. But although cats are solitary hunters, left to their own devices they do not usually live alone, at least if they are female. Recent research on groups of farm cats and feral cats has shown that females are surprisingly sociable.[2] They tend to live in small groups, often including mothers and daughters from previous litters. Within these groups, different litters may be reared in the same nest, with mothers suckling and caring for kittens that are not their own. But males really do lead quite solitary lives, and range over larger territories.[3]

There is a wide range of intensity in the relationships between cats and owners, and this helps to explain why cat-keeping is increasingly popular in many industrialized countries. Generally these relationships are fairly symmetrical. The more attention the owner pays to the cat's wishes, the more attention the cat pays to the owner. And since independence is so important to most cats, "acceptance of a cat's independent nature is one of the secrets of a harmonious human-cat relationship."[4] But cats can quite readily adjust to less interaction if their owners have little time for them or are not interested in forming closer bonds.

## Knowing when people are returning

Many cats seem to know when their owners are returning. I have received 615 accounts of this behavior from cat owners in response to my appeals for information. And in our random survey of nearly 1,200 households in Britain and America, we found 91 households with cats that seemed to know when their owners were coming home—in other words, about 8 percent of households have such cats.

About three-quarters of the reports I have received from cat owners concern an owner's return from work, shopping, or school, or from some other short absence. Here are some typical observations:

- "She is almost always at the window when I come home."
- "He appears from nowhere."
- "He is always waiting behind the door for us."
- "She is almost always there, and I wonder how she knows."
- And, from Ann Widdecombe, a well-known British politician, "No matter what time I come home, Pugwash is there at the door."

People who live alone usually do not know how long their cat has been waiting for them or if, indeed, it has been waiting there all day. Even when there are people at home, the anticipatory behavior of cats tends to be noticed less when the cats are free to roam outside. If the weather is good, some wait outside the house and are therefore less easily observed.

In 76 percent of the cases I know of, the cat waits for only one person, in 15 percent it waits for either of two people, and in 9 percent it waits for three or more. As with dogs and other animals, the people cats wait for are those to whom they are particularly attached, usually members of the immediate family or close friends. Here is a report sent to me by Jeanne Randolph of a cat living in Washington, D.C., that responded to two people:

> My boyfriend gave me a kitten named Sami for Christmas. Nearly every evening my boyfriend would stop by my apartment after work. I always knew when he was coming because Sami would sit by the door for approximately ten minutes before his arrival. I had no way of giving the cat signals because I was never aware of the time my boyfriend would be coming over. He was in real estate and worked odd hours. I doubt Sami could have heard his car as I live in the middle of a very noisy city in a high-rise. When my mother visits, she says Sami anticipates my arrival in the same fashion—and I take the subway.

In most cases where people paid attention to cats' waiting behavior, they found that the cats start waiting less than ten minutes before the person arrived. Nevertheless, practically all the stories involve behavior that does not seem explicable in terms of routine, familiar sounds, or other straightforward explanations. For example, when the son of Dr. Carlos Sarasola was living with him in his apartment in Buenos Aires, Argentina, he often came home late at night, after his father had gone to bed with their cat, Lennon. Dr. Sarasola noticed that Lennon would suddenly jump off the bed and go and wait by the front door ten to fifteen minutes before his son arrived home by taxi. Intrigued by this behavior, Dr. Sarasola made careful observations of the time the cat responded to see if the cat could be reacting to the sound of the taxi door shutting. He found that the cat responded well before the taxi arrived. "One night I paid attention to several taxis that stopped at the front of my building. Three taxis stopped and Lennon remained quiet with me in bed. Some time later, he jumped down and went to the door. Five minutes later I heard the taxi arrive in which my son was traveling."

Some cats make a point of meeting their owners on their way home from work or school, and a few even wait for them at bus stops or railway stations.

As with dogs, in some households the cat's behavior serves as a signal for someone to prepare food or make a cup of tea: "My father's cat went down to the front gate and sat on a stone gatepost waiting for him about ten minutes before he returned home," according to Joyce Collin-Smith. "As a journalist, his hours were very variable. My mother said that she knew to put the potatoes on when the cat looked up, apparently listened, then trotted off. It can't have been the distant sound of the car, however, because it went on even when he had no car and returned by bus and on foot."

In some cases a cat warns someone to break up an illicit party. That was the case with Bryan Roche:

> During my time as an undergraduate psychology student, I took a working holiday in Nantucket. The guesthouse in which I worked and boarded was inhabited by a Persian cat named Minu. Its owner (my employer) insisted that she had a psychic relationship with this cat and that when she was driving home the cat would growl for up to twenty minutes prior to her arrival. She often illuminated this fable with amusing recollections of her feline's psychic antics, and I regularly took to jesting with the residents about her unlikely stories.
>
> One night, however, unbeknownst to my absent employer, I held a small party in the guesthouse. When the party was in full swing I noticed that the cat was acting rather strange. She was arching her back, as cats do, but also growling quite loudly, like a dog. Given the gravity of being apprehended in the act of partying in my employer's house, I decided to heed the cat's warning and end the party. The guests were more amused by my superstition than by the cat's imitation of a dog. Sure enough, the cat's owner arrived home six or seven minutes later. The psychic cat had saved my job.
>
> I was still not convinced of the psychic nature of what had hap-

pened and I took to observing the cat very carefully. It quickly emerged that Minu could sense the arrival of her owner even when she arrived in a different car or at an unusual time. Her predictions proved reliable even when her owner was returning to the island from the mainland by boat! I became so convinced of the reliability of the cat's predictions that I held several more parties to which the cat was cordially invited. On each of these occasions, the cat proved to be a fail-safe employer-arriving alarm.

Although many cats respond to their owners' return on a regular basis, some do so only under certain conditions, most commonly when the owner's return is linked to their being fed. And some people have noticed that their female cats respond most when they are pregnant, but lose interest in their owner's return when they have kittens to attend to.

Of the 416 reports of anticipatory behavior on the database where the sex of the cat is given, there are almost equal numbers of responses by females and males, with 51 percent of the reports about females. In the random household surveys we carried out in England, slightly more females than males were said to respond: 52 percent as opposed to 48 percent. These differences are not statistically significant, and we can conclude that, on average, males and females behave very similarly in this respect.

## Keeping a log

Cats that are free to roam outside usually change their behavior according to the weather. On sunny days, they may wait outside in a sunny place near the door or gate; on rainy days indoors, on a windowsill looking out; and on cold days, somewhere warm.

This variability has so far frustrated the carrying out of videotaped experiments with cats, because if the camera is set up and left running pointing at a particular place, the cat may wait in another place, off camera. Dogs, by contrast, tend to wait in the same place, usually against the door or gate, and can be filmed more easily. To work effectively with cats would require a more sophisticated surveillance system than has yet been employed, or else the experiments would have to be restricted to cats that are kept indoors and always wait in the same place.

The behavior of cats that move freely indoors and out is more natural and more varied. It can be studied most simply and directly through the logs kept by cat-owning families.

The most detailed log so far is that kept by Judith Preston-Jones of Tonbridge, Kent, and her husband. Their two Siamese cats, Flora and Maia, usually reacted to her return after a short absence, while shopping or swimming, by waiting near the garage or on the doorstep. After longer absences, or in the evening, the cats anticipated her return by about ten minutes, waiting in a variety of places.

The log that she and her husband kept over a two-month period contains twenty-eight entries covering Judith's return at different times in the afternoon and evening. On fifteen occasions Mr. and Mrs. Preston-Jones went out together and so there was no one to observe the cats, but on all but one of these occasions the cats were waiting for them at one of their usual places on their return. The exception occurred when it was very cold and the cats were sitting on the boiler. On eight occasions Mr. Preston-Jones observed the cats showing signs of excitement and anticipation ten to fifteen minutes before his wife returned. Their waiting places varied according to the circumstances. When it was raining they waited indoors, either by the door or at the kitchen window; and when the weather was fine they waited in the garden, on the doorstep, or by the garage. On four occasions the cats were already outdoors with Mr. Preston-Jones in the garden and showed no special sign of anticipation. And at one homecoming the cats were nowhere to be seen; they were found hiding upstairs while a repairman worked on the washing machine.

The most interesting observation occurred one evening when Mrs. Preston-Jones came home at 9:40, following a meeting in a village church about three miles away. Her husband greeted her with "Well, the cats got it wrong this time! They got restless at nine o'clock, so I expected you home half an hour ago." In fact, she had left the church and gotten into her car, then remembered something she wanted to discuss with a friend. She had returned to the church and stayed there until 9:30. The cats reacted when she initially set off and got into the car.

## Aversions

Some cats anticipate the arrival of people to whom they have a strong aversion. Mosette Broderick, who lives in Manhattan, became the object of a cat's aversion by helping her former professor, who told her that his cat, Kitty, hated him for days after he took her to the vet. Mosette volunteered to take Kitty to the vet herself, so Kitty started hating her instead:

> As the years passed, Kitty developed her disgust toward me to such a degree that my professor always knew when I was on the block. When I turned down Sixty-second Street, from Lexington Avenue, some 200 feet and much noise away, Kitty would run and hide behind the stairs, which she did only when she expected my arrival. The curious fact here is that I would be out of hearing, sight, and smell range. In a crowded city like New York, she could not have heard me over the din of traffic. She certainly could not have seen me. Smell in winter in New York with the doors shut and the heat on in the house could also not have been a factor. I was also not always there on the same day or time, so schedule was not possible either.

At first, Kitty behaved like this only when Mosette was arriving to take her to the vet, but as time went on she hid before even the most innocent of visits.

## Cats compared with dogs

Fewer cats than dogs anticipate their owners' arrival. I have received only 615 cat stories compared to 1,133 dog stories. Of course these figures are only a rough guide, but a similar picture emerges from the random household surveys carried out in England and the United States. Out of a total of nearly 1,200 households surveyed, we found 91 households with cats that were said to know when someone was returning, and 177 with dogs that did so. The total number of dogs and cats in these surveys was practically the same. Overall, 55 percent of the dogs were said to show this anticipatory behavior, compared with 30 percent

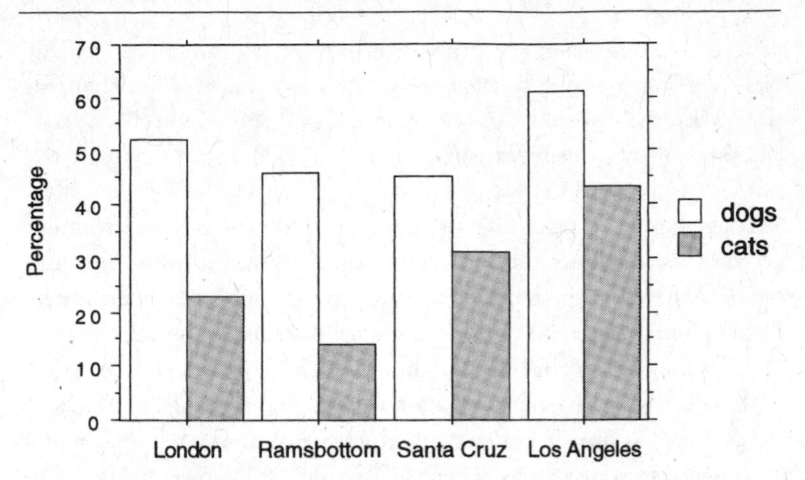


***Figure 3.1 The percentages of cat and dog owners who said their animals anticipated their returns. The surveys were carried out with a random sample of households in London; in Ramsbottom (in northwest England); and in Santa Cruz and Los Angeles, California.***

of cats. This difference between dogs and cats showed up in all four locations we surveyed: London and Greater Manchester in England, and Los Angeles and Santa Cruz in California (Figure 3.1).

The figures for anticipatory behavior by cats were higher in California than in England. I do not know why. Perhaps the Californian cat owners tend to form closer bonds with their animals than the English. But even in California, dogs significantly outperformed cats.

Are cats therefore less sensitive than dogs? Not necessarily. They may simply be less interested in their owners' comings and goings. And some may be only weakly bonded with the person returning. Nevertheless, many cats *are* interested in their owners' arrival and seem to anticipate their return.

The patterns of anticipation shown by dogs and cats also show characteristic differences. With dogs, a considerable proportion (17 percent) react when their owners are setting off to come home, or are intending to set off. With cats, this proportion is only about 1 percent. Also, with dogs a sizable percentage react when their owners reach a crucial stage in their journey, such as disembarking from a train or plane. Some cats do this too, but again, the proportion is very low, around 2 percent. Nearly all cats respond to their owners' return from work or shopping trips when their people are actually in transit. Why should this be so? I can think of two possible reasons:

1. Cats may be less sensitive than dogs, or for some other reason they may be unable to detect their owners' return until they are quite close to home. They may not be able to pick up their owners' intention from many miles away in the same way that many dogs seem to.
2. Cats may be capable of knowing when their owners are setting off, but they may have no motivation to respond very long in advance. If their aim is simply to meet and greet the returning person, there is no need to start waiting when the person is still far away. And while one of the traditional functions of dogs is to give warnings of people approaching, cats are not usually expected to play this role in the same way.

The way cats behave in anticipation of their owners' return from a relatively short absence seems compatible with both explanations. But the impressive way in which some cats respond to a person's return from a long absence suggests that they can be just as sensitive as dogs.

## Returns from vacations and long absences

Some cats show signs of anticipation hours before their people return from a long absence. If they have been kept by friends or neighbors, one of the commonest ways they do so is by returning to their own home. For example, Dr. Walther Natsch of Herrliberg, Switzerland, reported, "Our cat can sense when the family is coming home. While we were away the animal was with our neighbors. At the moment when we set off in Greece, Turkey, or Italy . . . the cat insisted on staying in our house again for the night."

Sometimes this behavior is unexpected and causes alarm to the person caring for the cat. "We went on holiday and left our cat with my aunt, just over two miles from our flat in the center of Brighton," said John Eyles. "When we returned two weeks later, the cat was sitting on the gate pillar waiting for us, and we were grateful to my aunt for having saved us the trouble of fetching him. When we rang to thank her, she was frantic; the cat had slipped away that very morning and she had been searching ever since."

Another way in which cats show this anticipation is by turning up to meet a member of the family who is returning for a visit. This is what happened to Elisabeth Bienz when she left her home in Switzerland to move to Paris, leaving behind her beloved cat, Moudi: "A few days later he disappeared from my parents' home and was not seen again. Every two or three months I came home for a visit, and the cat reappeared, well fed and cared for. My parents never learned where he was in the meantime. A few days after I had gone, he disappeared again. The biggest surprise came when I turned up for an unannounced visit one day. Some hours before my arrival the cat showed up. My mother was puzzled and thought he had made a mistake. But then I showed up too."

Similarly, Joan Forest, of Whidbey Island, Washington State, took over one of her sister's cats, which was shipped out by air freight from Boston, Massachusetts. When she let the cat out of the house, it vanished. She did all she could to find it but failed. Two months later her sister and her family came from Boston for a visit. "The day before they flew in, the cat came back! She was healthy and well fed. I have no idea where she was all that time."

On my database there are more than 150 examples of anticipatory behavior by cats prior to their people's return from vacation or a long absence. In most of them, as in these examples, the cats seemed to know of the impending return long in advance. And in some cases there is no possibility that they picked up this anticipation from the people who were caring for them.

Such cases refute the argument that cats have only a short-range awareness of an impending return. Their excitement and motivation are probably much enhanced after a long absence or vacation, especially when they have been taken away from their familiar surroundings. They are anticipating not only their favorite person's return but also a return to living in their own territory.

Although cats anticipate returns in a characteristically feline way, it seems clear that their anticipation cannot be explained simply in terms of routines and sensory clues. As in the case of dogs, it seems to be telepathic and depends on strong bonds between cat and person. I suggest that these bonds involve connections through morphic fields and that these are stretched, not broken, when a person goes away and leaves the cat behind. These bonds are the channels through which telepathic communication can occur, even over hundreds of miles.

Cats and dogs are not the only species kept as pets that anticipate the return of their people. As we will see in the next chapter, this ability is found in other species too, and even in humans. As with cats and dogs, this ability seems to depend on the formation of strong bonds that can act as channels for telepathy.

*Chapter 4*

# Parrots, Horses, and Other Animals

Among dogs and cats, the anticipation of their owners' arrivals depends on strong social bonds between the person and the animal. We would therefore not expect to find this telepathic ability in species that are inherently solitary, like most reptiles, or species that do not form strong bonds with humans, like stick insects. Even among species that are social and that do form strong bonds with people, it may well be that some are inherently insensitive to human feelings and intentions.

However, although much less information is available about species other than dogs and cats, there is enough to suggest that animals of at least twenty-seven other species also appear to anticipate people's return. Some humans do it too, especially in traditional rural societies.

Most species that anticipate their owners' returns are mammalian, but some birds also do so. Of the sixty-seven stories I have received about such birds, thirty-four concern parrots.

## Parrots

Parrots have the advantage over dogs that they can talk, and some of them announce their owner's arrival well in advance, like Suzie, a green Amazon who lived with the Lycett family in Warwick from 1927 to

1987. The father, a money collector for an installment plan company, used to travel to his collecting round in Coventry on a motorcycle. In the words of John Lycett: "As he did not have regular hours he could come home at all different times. My father's name was Cyril, which the parrot could not pronounce very well. In the evening the bird would be sitting quietly on her perch when suddenly she would get all excited and shout 'Werril,' and we would know that we could put the kettle on because my father would be home in half an hour."

Similarly, Deb Whitebread, of Missouri City, Texas, received advance warnings of her husband, Ron's, return from their African Gray, Rocket: "Ron taught Rocket to say *Hola* when he comes in the door. Then I noticed that Rocket would start saying *Hola* about ten minutes before Ron came home. It was not like he could hear his car or anything like that. Ron works a crazy schedule and comes home at different times. I used to have to call him to see when he was coming home, but now I just wait for Rocket to tell me. He does this on a regular basis."

Pepper was a youthful Amazon parrot who lived in Pennsylvania. He belonged to Dr. Karen Milstein and her husband, Philip, to whom the bird was closely bonded. "Our bird frequently starts calling 'Hello' and calling my husband by name shortly before he arrives home, even though the time may vary significantly from day to day," she told me in 1992. By 1994, when Pepper was seven, she noticed that he often reacted to her husband's intention to come home. In October 1994, Dr. Milstein kept a log, and here, for example, is the entry for October 17:

| | |
|---|---|
| 5:40 P.M. | Pepper is quiet |
| 6:14 | Pepper started calling "Hello" |
| 6:16 | Philip called to say he was leaving. Said he formed intent to leave two minutes earlier |
| | Pepper continued to call "Hello, Philip," until Phil came home just after 6:30 P.M. |

But although on most occasions Pepper became excited when Philip formed the intention to come home, and did so at nonroutine times, sometimes he did not respond at all until Philip pulled into the driveway.

Parrots can become very strongly attached to particular people and can show strong signs of jealousy, especially toward people of the opposite sex. Oscar, a blue-fronted parrot belonging to David and Celia Watson in Sussex, is strongly attached to David: "When he sees my husband, sometimes I can't go near him at all, he wants to attack me," Celia says. "I can't even touch his cage or give him his food. He is quite jealous. And he flings himself against the side of the cage when my husband leaves the room." Not surprisingly, Oscar gets very excited when David returns. His excitement begins ten to twenty minutes beforehand: "We thought he might be responding because David was coming home at a regular time, but it hasn't worked that way," Celia says. "With the job that my husband has now he never comes home at the same time, but Oscar is always waiting for him. He runs in his cage and starts fluttering his wings and makes little noises."

Most of the stories about parrots concern the owner's return from work or shopping, but some parrots reportedly react to a person's return from a longer absence. For example, when Peter Soldini went to France on vacation, he left his parrot with his mother in Switzerland, telling her that he intended to return in four weeks. Without informing her, he decided to come home after only three weeks, and he took three days on the return journey. "When I entered my mother's house the first thing she said was 'You won't believe how this bird has been acting these last three days. All day long he has been talking and singing. He is so excited.' "

## Other members of the parrot family

Other members of the parrot family also seem capable of anticipating their owners' return. I have received four accounts of budgerigars that show unmistakable signs of excitement five to ten minutes in advance, five of parakeets, and five of cockatiels.

Kathy Dougan lives in Santa Cruz, California, with six cockatiels. Friends who were in her apartment when she was out noticed that the birds became more active and chirped loudly before she returned. She kindly agreed to allow my colleague David Jay Brown to carry out ten experiments in which the birds were videotaped during her absence.

She returned at randomly selected times, when she received a signal through a beeper.

An analysis of these videotapes shows that on some occasions the birds chirped loudly when she was not on her way home—for example when the telephone rang or someone knocked on the door. But in seven out of the ten experiments the birds did indeed chirp more when she had set off to come home, a journey that took more than twenty minutes on foot. On average, over the whole series of experiments, they chirped loudly 15 percent of the time she was out and 49 percent of the time she was on her way home. These results were statistically significant.[1]

Birds that announce their owner's return by name, as some parrots do, are more likely to give unambiguous results than are those birds whose behavior is less specific, as is the excited chirping of cockatiels or budgerigars, but we have not yet found an opportunity to videotape an owner-announcing parrot.

Judging from the reports I have received, almost the only caged birds that anticipate their owner's arrival are members of the parrot family. This impression is confirmed by the random household surveys carried out in Britain and the United States. Thirty-eight of the households surveyed contained pet birds, four of which were said to anticipate the return of their owners: a parrot, a parakeet, a cockatoo, and a cockatiel. No finches, canaries, or other species were said to do so.

However, there is one exception to this generalization: a talking mynah bird called Sambo, belonging to the Rolfe family of Sutton St. Nicholas, Herefordshire (Figure 4.1). Sambo had a great rapport with the Rolfes' elder son, Robert, and used to tell the Rolfes when he was coming home from boarding school. "Two or three days before he was due to come home, Sambo would start chattering about Robbie," says Suzanne Rolfe. The family assumed this was because they had been mentioning his name more often than usual, but when he left school and started work, he was stationed in East Africa. "Sometimes he would let us know when he was coming on leave, but more often than not he'd arrive without warning. We always knew he was coming, though, because Sambo would start calling 'Robbie' a few days before he arrived."

*Figure 4.1 Suzanne Rolfe with the mynah bird Sambo (in the photograph against the cage) and Sambo's successor, Jacko, in Sutton St. Nicholas, Herefordshire. (Photograph: Phil Starling)*

## Chickens, geese, and an owl

A pet tawny owl called Joggeli lived with the Koepfler family in their flat in Zurich, Switzerland, for twenty-five years. In Heidi Koepfler's words,

> When something happened that was delightful to it the owl made a characteristic sound, a high-pitched *rr-rrr-rrrr,* like a bell. At the

> same time it closed its eyes. When our sons came home from school or university we always heard Joggeli's joyful sound when it could not see or hear them yet. My brother, who lives at another place and rarely visits us, did not take Joggeli seriously and laughed about the bird. One day Joggeli made angry, aggressive sounds and flew against a windowpane. I thought, "What is wrong with it? It does this only when Ralph comes." And really, my brother paid us a surprise visit.

Some chickens react to the person who feeds them. For example, when Roberto Hohrein was in school in Germany his family kept ten chickens. It was his job to feed them when he came home from school. His mother found that ten to fifteen minutes before he arrived, they seemed to be waiting for him, standing in the corner of their chicken run from which they could see him as he approached. "What surprised my mother was that they did not stand there at the same time every day but at different times, according to my timetable. German schools do not finish at the same time every day. Sometimes I did not come by public transport but found a car to pick me up. But no matter when it was, the hens always stood there and waited for me because they were hungry. Only when I came extraordinarily early they did not pay attention. That was when they were not hungry yet."

If chickens are less motivated by personal attachment than by a desire to be fed, one story about geese suggests that a bond with a particular person was their principal motive. Herr K. Theiler, living near Thun, in Switzerland, had three pet geese with whom he had a particularly close relationship: "Even my mood, happiness or sadness, was reflected in their behavior." His wife was able to tell from them when he would be coming home from his office. "The geese waited impatiently at the entrance to the garden. Usually I was home at twelve-fifteen, but if something had come in between she saw that I would be late because the geese were quiet."

Birds from a wide range of species are known to form strong attachments to people, especially if they have been raised by them from an early age,[2] and it may well be that other species are capable of this kind of anticipatory behavior, over and above those I have already heard about.

The wild ancestors of most domesticated birds, including geese, chickens, and birds of the parrot family, lived in flocks. Maybe their ability to anticipate the arrival of a human companion is derived from an ability to know when separated members of the flock are approaching. Or perhaps it is more related to an ability of young birds to know when their parents are returning to the nest with food. But nothing seems to be known about this kind of anticipatory behavior in the wild.

If further investigations of pet birds confirm that members of some species can indeed anticipate the arrival of their owners through a kind of telepathy, then it would be worth observing birds in the wild. Do they seem to anticipate the return of other birds with whom they are closely bonded? Do young birds in the nest anticipate the arrival of their parents with food?

It should also be possible to do experiments with domesticated birds, such as geese, to see if they can anticipate the return of a bird that has been taken away beyond the range of sight and hearing and is then brought back or allowed to return by itself. Experiments should also be possible with homing pigeons. Do birds left behind in the dovecote show any signs of anticipation before their mates and other companions return from a race?

## Reptiles and fish

In the wild, most reptiles are solitary, coming together only to mate. Moreover, in most species when the females have laid their eggs, they abandon them, and the young have to fend for themselves. Think, for example, of baby sea turtles hatching on beaches thousands of miles from their ancestral feeding grounds, which they have to find on their own with no adults to guide them. And if wild reptiles do not form strong bonds with each other, captive reptiles will have little inherent capacity to form bonds with their human keepers.

These negative conclusions about reptiles are reinforced by one of the most experienced of my correspondents, Jeremy Wood-Anderson, a naturalist and reptile collector who lived in Pakistan where he kept a wide variety of reptiles for more than thirty years. While he was convinced that "psychicness" exists to varying degrees among mammals and birds, he did not think it exists in reptiles in any recognizable form. He was convinced that they cannot pick up their owners' thoughts telepathically:

"Beyond reactions to habits they have got used to, there is absolutely no connection between the mental processes of reptiles and humans."

In the 1990s I made sustained attempts, including appeals for information in specialist publications like *Reptilian International*, to find out if any people who keep reptiles as pets had noticed return-anticipating behavior. I came across no convincing examples, as I described in the first edition of this book. However, several people who read my negative conclusions wrote to tell me of experiences that suggest some reptiles can indeed tell when people are coming. For example, Hilary Lennox, of San Francisco, formed a close connection with a boa constrictor named Julian, with which she danced.

> For several years I borrowed Julian from Charlie, a vet who practiced in Berkeley. I went to Charlie's office regularly to get Julian, and I kept her for a week or two and used her in my belly-dancing performances. Julian spent many hours curled up next to me on my bed while I was studying and seemed to enjoy being taken out to participate in my dancing and visit with people in the audience. I always felt she and I had a mutual relationship based on affection. The secretary in Charlie's office told me that every time I came to pick her up, about half an hour before I arrived Julian awoke from her deep snake sleep and started moving around her cage, sliding her nose back and forth across the cage door.

Jane Schooley, of Whitehall, Pennsylvania, kept nine lizards of different types and noticed that her iguana "seems to know that my husband is coming up the road long before it is possible to see or hear him."

Mrs. D. Delaney, who lives in Essex, England, noticed that her tortoise, Fred, seemed to know when her granddaughter was coming to visit. "I bought him when my granddaughter was four years old, and she is now twenty-three. No matter what time she comes home in daylight, he walks to the patio doors, waiting for her, and whenever she is in the garden he follows her."

Thus it seems that some reptiles can form bonds with people and anticipate when they are coming home, though this appears to be rare. Relatively few people keep pet frogs, newts, and other amphibians, and there are no reports to suggest that they form psychic bonds with humans or respond to them telepathically. The same goes for pet insects, such as stick insects.

Many fish species are more social than reptiles or amphibians. They swim in schools, or shoals. And some species, including some kinds of cichlids popular with keepers of tropical fish, build nests and protect the eggs and the fry. But even in species where there is some degree of parental care there is little scope for human beings to substitute for fish parents and form bonds with the young.

The keeping of fish is far more common than the keeping of reptiles, amphibians, and insects. In the United States, there are about 76 million pet fish, and about 14 percent of households keep fish.[3] There are plenty of opportunities for people to notice whether their fish get excited before a particular member of the family comes home. But I have not heard of a single instance or observed anything of this kind with our own family goldfish, nor have I found any evidence for any other kind of telepathic connections between people and fish.

## Guinea pigs, ferrets, and other small mammals

By far the commonest mammalian pets are dogs and cats, but a variety of other species are quite widely kept, including rabbits, guinea pigs, rats, mice, gerbils, hamsters, and ferrets. I have received no reports at all of psychic gerbils, hamsters, rats, or mice. I have received only one inconclusive report about a house rabbit, one about a pet rat, and five about guinea pigs, but none is said to have reacted more than two or three minutes beforehand, and it is impossible to rule out the possibility that they were reacting to familiar sounds.

Of all the small mammals I have heard about, the only one that seems promising from a telepathic point of view is a ferret in the East End of London. This animal is strongly bonded with its owner, while the owner's wife and the ferret share a mutual dislike. Joan Brown has noticed that the ferret waits for her husband at the front door before he arrives home: "If the ferret is in the living room, she either hears the car before I can or she somehow knows my husband is on his way, because she races to the door a good ten minutes before he arrives. Sometimes he will come home late but she still knows. If he stops off for a drink with co-workers he could be an hour late, but she will be waiting for him an hour later than normal too."

## Monkeys

John Bate, of Blackheath, South London, had a return-anticipating squirrel monkey:

> When I was commuting between Coventry and Blackheath, the monkey would let my wife know when I was north of the Blackwall Tunnel [under the river Thames] by chuckling in a distinctive manner. Entertaining a friend one Friday afternoon my wife announced that I would be home within fifteen minutes. "How do you know?" asked her friend. "The monkey has just told me," replied my wife. Within a quarter of an hour they heard my key in the door. However acute and discriminating the hearing of an animal, it seems doubtful that it would be possible to distinguish one car from another in heavy London traffic, four or five miles away, either across or under the water of the river Thames.

I agree with this conclusion. I have heard of several other return-anticipating monkeys, but they are so rarely kept as pets nowadays that there is little scope for further research with them, fascinating though this would be.

## Horses

Together with dogs, cats, and parrots, horses are the nonhuman species with which people form the strongest relationships. Many riders feel themselves closely bonded with their horses, and some are convinced of a psychic link. I discuss more general evidence for human-horse telepathy in Chapter 8. Here I am concerned specifically with the ability of horses to know when their owners are coming home.

Many people have found that their horses seem to know when they are approaching the stable. They may become more alert, show signs of excitement, or whinny. But most owners are not sure how long in advance the horse responds, to what extent its reactions are a matter of routine, or whether the response is attributable to sharp hearing. Moreover, since horses do not live in houses, they are usually less closely observed than dogs, cats, and other house animals.

Those who have the best opportunities to notice anticipatory behavior in horses are people who work in stables or who care for other people's horses while they are away.

When Adele McCormick and her family went away from their ranch near Calistoga, California, they usually left their thirteen horses in the care of people who knew when they would be returning. Their horses did indeed seem to anticipate their return, but they could have picked up this anticipation from the people looking after them. On one occasion, however, they were looked after by a stranger who did not know when the family would be coming back: "When we arrived home the man greeted us and said, 'I knew you were on your way home because the horses started acting strange.' He said that while he was feeding them, 'Instead of looking at the food like they normally do, all thirteen horses kept looking down the road, running and whinnying.' He said this started at 4:30 P.M. We arrived at the ranch between 5:15 and 5:30 P.M."

Sometimes horses show anticipatory behavior hours in advance, especially when their person has been away for a long time. This happened over and over again with Elliott Abhau who, because of her work, had to leave her two cherished horses with her best friends on a farm in Maryland. She came to visit every few weeks at irregular intervals over the next ten years. She did not usually tell her friends when she was coming, but they told her that they always knew because of the way the horses behaved: "The day before, they pick on each other, which never happens otherwise, and then, on the day, stand at the fence together looking down the driveway." This happened hours before her arrival by car, the journey taking four to six hours.

Herminia Denot grew up on a ranch in Argentina and learned to ride almost before she could walk. She was very attached to her horse, Pampero, but she had to leave him behind when she went to high school in Buenos Aires, returning to the family ranch for her vacations. The gaucho who looked after her horse noticed that as the time of her return drew near, "Pampero would go crazy. He galloped around the ring neighing." The day before her arrival he would stop in front of the corral gate, looking north, in the direction of the train station. But on one occasion her parents brought Herminia home by car, and this time Pampero surprised the gaucho by looking to the southeast, not to the north where the trains ran. The direction he was looking in was in fact the direction from which she was approaching by car.

Finally, an English example: Fiona Fowler got her New Forest pony, Joey, when she was twelve and broke him in herself. When she went away to study nursing in London, she had to leave Joey with her mother near Winchester. She went home on days off about twice a month. Her mother noticed that Joey always seemed to know when she was on her way home; he made his way from a lower paddock, where he spent most of his time with other horses, and waited at the gate. He continued to do this over a period of years whenever she was returning. "There was one particular occasion when I wasn't expected home and my mother was surprised to find Joey waiting at the gate as usual. Ten minutes later I phoned from the station requesting to be picked up."

Stories like these show that some horses seem to know in a seemingly telepathic way when their owners are coming. The next stage in this research would be to carry out videotaped experiments with such a horse, recording its behavior while the owner sets off for home at randomly selected times.

## Sheep and cows

Sheep are not often kept as pets, but when lambs are raised by people they can form close attachments, as in the nursery rhyme "Mary Had a Little Lamb."

Margaret Railton Edwards and her husband, Richard, found themselves the owners of a lamb when some sheep-farming friends left a sick one, still on the bottle, at their home in Cheshire. The Edwardses nursed the lamb back to health, and he lived with them in the house for about four months. "Shambles was almost house-trained," Margaret Edwards said, "and would sit on my knee watching TV in the evening. My husband, Richard, would come home between five and seven P.M. About ten minutes prior to his arrival Shambles would sit by the front door and wait for him. Even if Richard was in a friend's car he would still wait at the door. Occasionally Richard would come home at lunchtime, and the same thing would happen."

I have heard from two other people who kept pet sheep who had similar experiences. One lamb, Augustus, was adopted by the Ferrier family on Whidbey Island, Washington, and formed a particularly strong attachment to Grant, then age fourteen, who fed him, took him for walks, and played ball games with him. Grant's father, Malcolm, told me that Grant came home in the afternoon at irregular times because of after-school activities. But the family could always tell when he was on his way: "Augustus perked up, baaed, ran about his pen, and showed every sign of an event about to happen. And then five minutes later Grant and his pals would appear." Could Augustus have known by any normal means? Malcolm Ferrier does not think so:

> We talked often about his sequence, and were completely convinced that there was no normal physical method by which Augustus could be aware of Grant being on his way. He couldn't see him (much vegetation), or hear him when he started his welcoming routine, particularly over the suburban traffic noise. It was very clear to us all, in an amateur and nonexperimental way, that some sort of strange communication was taking place; the neighbors often remarked upon it too. Grant often tried, unsuccessfully, to creep up on the beast.

I have come across only two examples of return-anticipating cows. One concerned a nun, Sister Veronica, at a convent in Coolgardie, Australia, where she worked in the convent kitchen and was known for her outstanding ability to handle the cows that supplied the convent with milk. When she was returning from her vacation, on the morning she was expected home, the cows were seen to go and stand near the railway station where she would arrive.[4]

These stories about sheep and cows, though few in number, agree well with the pattern of behavior shown by dogs, cats, horses, parrots, and other animals. The ability to anticipate a person's return seems to occur in a wide range of species. In every case, it seems to depend on the formation of close bonds between the person and the animal.

Those species that do not show this kind of anticipation, such as fish, may not do so either because they are inherently insensitive to telepathic influences or because they are incapable of forming bonds with people who are strong enough to act as channels for telepathic communication.

Presumably this ability has not evolved simply in the context of pet-keeping but also occurs between animals in the wild. I return to a discussion of animal-to-animal telepathy in Chapter 9.

If anticipation of returns is so widespread among nonhuman animals, we might expect some people to have a capacity to know when other people are about to arrive.

## Humans

Stories abound from people who have lived or traveled in Africa about the way in which some Africans can anticipate arrivals in the absence of any known means of communication. For example, Laurens van der Post found that Bushmen in the Kalahari Desert of South Africa could tell when members of their group had killed an eland 50 miles away from their camp and when they would be returning. The Bushmen who hunted it were traveling with van der Post, and as they drove back toward the camp in Land Rovers filled with meat, van der Post wondered how the people there would react when they learned of their success in hunting. One of the Bushmen replied, "They already know." Sure enough, as they approached the camp they heard the song that was used to celebrate on such occasions. When the eland was killed, they immediately knew "by wire," as the Bushman put it. Van der Post found that they were "evidently under the impression that the white man's telegraph also worked by telepathy."[5]

Many people familiar with Africa have had similar experiences. A young European man named Sinel who lived among the tribesmen of the southern Sudan remarked that "telepathy is constant." They always knew where he was and what he was doing even when he was far away. On one occasion when he got lost, men came out to collect him, as if they had sensed his plight. On another, when he had picked up an arrowhead and brought it back with him, two tribesmen came to ask him if they could examine it.[6]

Probably such abilities were better developed in traditional societies than in the modern industrial world. Even in some parts of Europe they seem to have been widely recognized. The "second sight" of the Celtic inhabitants of the Scottish Highlands included "visions of 'arrivals' of persons remote at the moment, which later do arrive."[7] In Norway there is even a special name for the phenomenon, *vardøger,* which literally means "warning soul." Typically, someone at home hears a person walking or driving up to the house, coming in, and hanging up his coat. Yet nobody is there. Some ten to thirty minutes later, similar sounds are heard again, but this time the person really arrives. "People get used to it. Housewives put the kettle on when the *vardøger* arrives, knowing that their husband will arrive soon."

Professor Georg Hygen of Oslo investigated dozens of recent cases and concluded that this phenomenon is more telepathic than precognitive—in other words, the *vardøger* is not so much a pre-echo of what will happen in the future as something related to a person's intentions. For one thing, the sounds are not always identical to those heard in advance. A person might be heard going up to the bedroom whereas when he arrives he goes to the kitchen.[8]

The *vardøger* phenomenon can occur, however, when the person does not arrive, having changed his mind. One man, for example, arranged to meet his wife in a store. He then decided to pick her up at her office instead, but he was unable to do so because he was delayed, so he went to wait for her at the store as originally planned. She did not arrive, and after waiting for an hour he went home. When she came home, she complained that he had not come to her office. She had heard his *vardøger* and, from previous experience, trusted it so much that she waited at the office for an hour before she gave up.

In English we have no synonym for *vardøger.* Nevertheless, some people have noticed and commented on anticipated arrivals, although none have mentioned the sound effects typical in Scandinavia. Of these accounts, most concern parents and children, and the remainder involve husbands and wives. In some cases children seem to anticipate a parent's arrival. Belinda Price provided this example of a baby doing so:

> Until my son was about eight months old I always knew when his father was on his way home. About seven or eight minutes before he

> arrived my baby would become very alert and then expectant. As we lived on an active air base at the time I don't think he heard anything, and my husband rode a bicycle some of the time. He used to return home unexpectedly at any time of day or night, as he was a pilot used to scrambling [making emergency flights].

Other parents have told me that when they go out in the evening leaving their baby with a sitter, quite often the baby wakes up shortly before they arrive home. And when children are old enough to talk, some actually announce the imminent arrival of a parent. This happened when Sheila Michaels was looking after a three-year-old boy in New York while his mother was hospitalized. "I did not expect his mother to be released for another day. I was reading a favorite story to him when the boy got off the bed and went to the door, calmly saying, 'Mommy, Mommy,' but in a way that upset me terribly. I tried to get him to come back and read the book with me, but he could not be budged, repeating, 'Mommy, Mommy,' endlessly. I told him she would be back tomorrow and his father was due in a couple of hours. He was immovable. Then his mother walked in."

I have heard of no cases of fathers anticipating the return of their children, but several of mothers. Here is a dramatic example from World War II, sent to me by Mrs. C. W. Lawrence: "During the war my brother Jack served in the Royal Navy and during active service was not allowed to write home. One evening when Jack had been away for more than two years, my mother suddenly stood up and said, 'I must get Jack's bed ready. He'll be here tonight.' 'What on earth makes you think that?' we asked, laughing at her. 'I just *know* he'll be here,' she said and went upstairs to attend to the bed. Later that evening Jack arrived!"

Most instances of anticipation are more mundane. Bonnie Hardy, a mother of teenage boys who lived in Victoria, British Columbia, found this phenomenon wore her out through lack of sleep. When the older boys returned home very late on weekends, despite their efforts to be as quiet as possible, she was still being disturbed. "Nothing worked, and then I realized that it was not just their movements in the house that disturbed my sleep but the fact that I woke up when they got in their car to head home." She woke up first, and then heard them arrive.

In a similar way, some women find they wake up before their husbands come home. Cindy Armitage Dannaker, who lived in Pennsylvania, was one of them: "It has happened so often that now I just say to myself 'He's coming,' and wait. Usually within five minutes or so I hear my husband's Jeep pulling up our road. I feel like he thinks of me or something and I pick this up in my sleep. All I know is that suddenly for no apparent reason I am wide awake and I feel he is coming."

Sometimes this anticipation occurs well in advance, particularly when people have been separated for a long period. And sometimes people act on these feelings in an appropriate way—for example by going to meet a particular train or plane.[9] Here is a particularly striking account recorded by Oliver Knowles:

> I was employed by the United Nations for fourteen years, during which time I had to do much traveling. But on only one occasion did I return early to Geneva because of illness, during the 1970s when I was in Abidjan. I did not inform my wife that I was returning, as I didn't want her to fuss, and she was at the time on holiday in Austria with our four sons. However, when I arrived back in Geneva she was waiting for me at the airport. She said that she had had an overpowering feeling that she must meet that particular flight so had packed up the family and returned.[10]

If such cases of anticipation by people are viewed in isolation, they seem like scattered anomalies. But in the context of anticipatory behavior by a wide variety of animal species, they fit into a larger pattern. The anticipation of arrivals seems to be an important aspect of the natural history of telepathy. The fact that these instances of anticipation can occur in babies and when people are asleep shows that they are not dependent on the higher mental faculties. They work at a more fundamental level and are rooted in our long biological and evolutionary heritage.

# Part III

# Animal Empathy

Chapter 5

# Animals That Comfort and Heal

The word "empathy" means "a sympathetic understanding or suffering."[1] As we have seen, it shares the Greek root *pathe* (feeling, or suffering) with "sympathy" and "telepathy." However, I am not suggesting that empathy and telepathy are necessarily linked. People no doubt pick up other people's feelings through body language and other sensory information, and as I demonstrate in this chapter, animals are sensitive to people in the same way. What is of interest here is not so much the way the feelings are transmitted as the fact that the animal responds to them so sympathetically.

Mutual help is an essential aspect of social life in many animal species. Even those who believe all animal behavior is shaped by "selfish genes"[2] acknowledge the importance of altruistic behavior in ant colonies, in parental care in birds and mammals, and in social groups of every kind.[3] For example, when a member of a herd or flock gives an alarm signal, alerting other members of the group to danger, it may be endangering itself by drawing the attention of a predator toward it.[4]

The selfish-gene theorists acknowledge the reality of altruism in animal social groups, but explain it in terms of selfish genes working for their own survival and reproduction. An individual animal may lay down its life for the greater good of the genes it shares with its offspring and close relatives.

Altruism between pets and human beings cannot be explained in terms of selfish genes in any straightforward way. A person helping a sick pet, caring for it, and paying veterinary bills, is behaving altruistically, but not because of selfish genes shared by the pet and the person. Pets and people have very different genes; they belong to different species. And just as people help pets, so pets help people, not least through their emotional bonding.[5] People form the closest bonds with species that show the greatest empathy toward them—above all, dogs, cats, horses, and parrots.

## Keeping pets can keep us well

Our own cat was called Remedy because my wife, Jill, soon found that she was just that. Her warm, purring presence was indeed a remedy. She seemed to sense when she was really needed, and at those times she would sit or lie on Jill or me, working her healing magic.

On my database there are more than six hundred stories about animals that comfort and heal. Most of them are about cats and dogs staying close to people who are sick or sorrowful, as if to comfort them. Indeed, there is no "as if" about it. They *do* comfort people and even help to heal them. A number of scientific research projects have even quantified their beneficial effect.

For example, in a study carried out in the Philadelphia area, elderly people who adopted cats were compared with a similar group of elderly people who did not adopt cats. Regular follow-up interviews and tests showed that within a year there were striking differences between the two groups. As measured by standard psychological tests, the cat owners felt better, while the nonowners felt worse. And although the owners and nonowners did not differ significantly to start with, after a year those with cats felt less lonely, less anxious, and less depressed. The cats also had a favorable effect in reducing blood pressure in people with hypertension, and in reducing the need for medication.[6]

Of course, these benefits depended on the bond that developed between the person and the cat. Companion cats provided fun, company, and affection and helped take people's minds off their troubles and their ailments. The stronger the bond, the greater the positive effects seemed to be.[7]

Likewise, relationships with dogs can reduce blood pressure and confer other physiological benefits.[8] These benefits may also be experienced by the dogs themselves, as their heart rates drop while they are being petted.[9]

In a study by Erika Friedmann and her co-workers at the University of Pennsylvania, pet owners who had been hospitalized with heart disease, including heart attacks, showed a higher survival rate a year later than a control group of non–pet owners.[10] The presence of a pet at home was an even stronger predictor of survival than having a spouse or extensive family support.

Pets can also help people who are bereaved. Several studies of people who have recently lost a spouse have shown that pet owners were less depressed and less prone to feelings of despair and isolation. They also had better general health and needed less medication.[11]

But it is not only sick, elderly, bereaved, and vulnerable people who can benefit from keeping pets. These effects are quite general, both for adults and for children.[12] Dogs in particular help people make friends. And research by James Serpell at Cambridge University showed that most people who had recently acquired dogs developed a greater sense of security and self-esteem. Their general health improved, partly because of the increased amount of exercise they took in walking the dog. They also suffered less from minor ailments like headaches, colds, and flu.[13]

Pets can offer children both companionship and security, responding to demands and giving uncritical sympathy. Disturbed children seem to rely especially heavily on animals as a source of support. In one study delinquent adolescents were twice as likely to talk to their pets and three times as likely to seek out their animal's company when lonely or bored.[14] Pets can also help children to develop a better sense of mutuality and responsibility as they care for the pets and respond to their needs.[15] And there is evidence that pets can help smooth relationships and improve family dynamics. For example, in a study in Baltimore, Maryland, of the impact of pets in sixty families, many of the families experienced increased closeness, spent more time playing together, and argued less after they obtained their pets.[16] A review of NIH-funded research in 2009 shows a wide range of benefits of pet ownership on both psychological and physical health.[17]

Although most studies have reinforced the message that pets are good for you, this is not always the case.[18] Pets are not magical. They are good, bad, or indifferent, like people. And some people acquire a pet precisely because they want benefits from it, and the heavy weight of expectation can lead to the animal being unceremoniously abandoned or killed because it develops behavior problems or fails to make the owner feel better.

In addition to the dogs that are kept simply as pets, there are many that help people in very practical ways, including sheepdogs and other working dogs and the service dogs that play a vital role in the lives of many thousands of people. The best known are guide dogs for blind people, but there are also hearing dogs for deaf people, dogs that assist people with disabilities, and dogs that alert epileptics to oncoming seizures.

There are also many programs—more than two thousand in the United States alone—in which animals visit people in hospitals, hospices, and homes for the elderly. These animals usually belong to volunteers and are often called PAT (pet as therapy) animals. They are helpful for children, especially for the chronically sick, many of whom eagerly await their animal visitors.[19] They also are very popular among elderly people and among people in hospices, where they can have a relaxing effect on both patients and staff, lighten the mood, provide affection and physical contact, and act as social lubricants.[20]

Prisons that allow animals to visit prisoners or allow prisoners to keep pets themselves have seen a reduction in violence, suicides, and drug use as well as improved relations between prisoners and staff.[21]

How is it that animals can be so beneficial to humans? Attempts to explain their influence include words like "empathy," "acceptance," "companionship," "emotional security," and "affection." These are the same kinds of phrases that are often applied to the healing effects of other people. The secret of this healing power seems to be the same whether it comes from people or from animals: unconditional love.

Loving unconditionally seems to come more easily to dogs and cats than it does to most human beings. The loving behavior of pets is both a cause and an effect of the bonds they form with people. It is expressed most notably when their owners are in need.

## Comforting cats

One of the most consistent features of accounts of the comforting and healing behavior of companion animals is that they respond to people's needs. They are not simply behaving in a generically affectionate manner. For example, Jahala Johnson of Antioch, Tennessee, reports that her cat "always seems to know when I need comfort. One night as I lay down for bed after a very stressful day with the world's troubles heavy on my mind, Kitty jumped up on me, ran up my chest, meowed and placed her paw gently on my face. She seemed to say, 'It's okay, Mom. I love you.' Then she snuggled up under my chin. That was the best medicine I could ever have."

The responsiveness of cats is especially striking in animals that normally cherish their independence. Gertrude Bositschnick of Leoben, Austria, writes that "For fifteen years Baerli, a yellow male cat, was my loyal companion, the joy of my life. He was a gorgeous cat who loved his freedom. When I did not feel well or was sad, though, he never parted with me. Instead he lay on my lap purring and pressing himself close to me. When I was well again he was off as usual, especially at night."

When there is more than one cat in the household they sometimes take turns. Karen Richards, of Stourbridge, in the West Midlands, lives with five cats, and when she was very unwell for months on end, unable to go to work, one of the cats stayed close while the others roamed free. She said that the cats had a rotating schedule for going out, so that she was never left completely alone in the house.

Several people have reported that their cats comforted them when they were grieving over the death of a loved one. For example, Murielle Cahen of Paris said, "Both cats stuck with me as if they didn't want to leave me alone with my sorrow, and this lasted exactly the time I was mourning. After that the cats were more aloof again."

Many people have commented that their cats behave in an unusually considerate manner when they are ill. A common feature of these stories is that this considerate and comforting behavior by cats happens when needed and goes on as long as necessary. But when the person has cheered up, calmed down, or gotten better, the cat reverts to its usual independent behavior.

## Devoted dogs

Many dogs, like many cats, seem to sense when their people need comfort. Jeanette Hamilton, of Redwood City, California, for example, found that her Standard Poodle, Marcus, was extremely sensitive to her emotions. "Whenever I cry silently he comes to me and licks away my tears. He tunes in whether he's at my feet or in another room, whether he's asleep or awake."

Out of more than 300 accounts of such behavior by dogs there are many comments like these: "My dog senses exactly when I do not feel well or am sad" and "When I am sad, she does not leave me and puts her head on my knees." One of the simplest yet most eloquent is from Sue Norris, of St. Helens, Lancashire: "I am autistic and have a dog Nickita, she knows how I am. She comforts me before I have told her. Sometimes I have bad days. She is there with me where I am."

Many dogs also seem to know when their person is ill and behave very considerately, staying close and behaving in a truly comforting way. Rosemarie von der Heyde, of Achern, Germany, had a Dachshund who usually greeted her enthusiastically on her return home. "But once I had injured my heel and when I came home he reacted very differently. He just stood there without moving and looked at me. Slowly he came to me and held out his paw. I lay down on the sofa and, contrary to his normal behavior, he did not start jumping on me. He quietly lay down next to me as if to console me."

Sometimes dogs also seem to know what part of the person's body is painful, and comfort them where it is needed. John Northwood, of Poole, Dorset, was a retired policeman who believed dogs should not be allowed on the beds. He often took his daughter's Collie, Ben, for walks, but on one occasion he had a bad back and had to lie down. "As my head hit the pillow," he recalled "the bedroom door opened and in came Ben. He jumped up onto the bed and stretched out against my back. I felt too ill to say anything, but the feel of him against my back was good. He must have sensed that I was unwell and needed warmth."

Some dogs lie with their people while they are suffering from migraines. Frau R. Huber, of Horgen, Switzerland, found that her dog, Nero, also knew on which side of her head she had the migraine. "If it was the right side he excitedly and vigorously licked my right eye and my right forehead with a low whimper. If the pain was on the left he did the same on the other side. It was like a massage."

## Animals preventing suicide

As we have seen, both dogs and cats can be very sensitive to their people's moods and emotions. In some cases their responses go beyond comforting; they can literally save people's lives.

In the midst of a stressful marital problem, a woman in the north of England decided to end her life. Leaving her dog and cats "sleeping contentedly in a pile in front of the fire," she went into the kitchen for water and acetaminophen tablets. Suddenly William, her beloved English Springer Spaniel, jumped up, ran in front of her, and for the first time in the fifteen years of his life, "He snarled! His lips were pulled completely back so that he was almost unrecognizable," she says. "Horrified, I replaced the bottle top and, genuinely afraid of the dog, I went back into the room and sat on the sofa. William bounded after me, leaped on to me and began frantically licking my face, his whole body wagging."

In some cases dogs have prevented a suicide by alerting others. A German dog called Rexina was shut in the house one day by her owner while he went to a shed in the garden. The dog waited by the door, but after a while she howled and came running to the other members of the family. "She was very excited," Dagmar Schneider said, "and we noticed that our father had been gone for quite a while. We let her out and looked for him. When we found him he said, 'Thank God you came!' Later he admitted that he had intended to commit suicide. Rexina had felt it, and if she had not been there we would have been too late."

Cats too have stopped people from killing themselves. P. Broccard tells of a Swiss cat called Pamponette. "I was feeling really low and wanted to kill myself. My cat must have felt the state I was in. That day she did not leave my side for one moment. She, who normally never meows, meowed all day, and she rubbed her head against mine each time I sat down. In the afternoon Pamponette usually slept with my other four cats, but she never left my side, and during the night she slept next to my pillow, where she does not normally like to stay."

Her behavior was very like that of cats who comfort their owners when they are sick or upset, but here the stakes were higher.

## Animals as therapists

The ancient Greeks thought that dogs could cure illness, and kept them as co-therapists in their healing temples. Asklepios, the chief healing divinity, extended his power through sacred dogs.[22] Although they have no such acknowledged part to play in modern medicine, in practice they have found their way back into a healing role through pet-as-therapy programs run by volunteers.[23]

Some effects of animals taken to visit sick or elderly people are generic: They have a comforting and cheering influence and take people "out of themselves." But sometimes the animals show a remarkable sensitivity to the needs and condition of a particular person. For example, Chad, a Golden Retriever, goes almost every day with his owner, Ruth Beale, to visit a hospice in Birmingham, England. "He seems to know which patients are really ill, compared to the others, whom he acts the clown with," Mrs. Beale says. "He will just sit there with his head on the person's lap or on the bed, or he'll stand there quietly with them. There was one particular lady he became very close to, and we had a telephone call at 10:00 P.M. saying she was dying and she wanted Chad with her. And he stood with his head on the bed for three hours while she died." Chad won the PAT Dog of the Year award in 1997 for his hospice work.

Deena Metzger used to keep a wolf called Timber when she was working as a counselor and living in the country near Santa Monica, California. Timber showed a remarkable sensitivity. "I watched him discern my patients' needs and come to them, laying his head quietly in their lap, when they were experiencing pain too great to be comforted by a human. His intuition was infallible."[24] Other counselors and therapists have also found that their dogs or cats can be very perceptive about their patients' needs and even act as co-therapists. Even Sigmund Freud was assisted by his dog, a Chow, who was no mere ornament but part of the process, the "petting cure," as he called it. She would "sit quietly at the foot of the couch during the analytic hour." But toward the end of the session, she helped Freud by "unfailingly beginning to stir," showing that time was up.[25]

Horses have a remarkably therapeutic effect on people with mental or physical problems, including people with Down's syndrome. For many years there have been riding programs in Britain and other countries for people with disabilities, who can gain a new confidence and sense of freedom. These can result not only in psychological benefits but also in improved balance and coordination.[26]

Near San Antonio, Texas, Adele and Deborah McCormick, a mother and daughter team, work as therapists with people with serious mental illness, criminal behavior, and drug addiction. But their work as psychotherapists took on a new dimension when they enlisted the horses on their ranch in the healing process. "The size, strength, and physical presence of the horse make people more aware, quite literally bringing them to their senses. . . . Equine therapy is for anyone who feels down, demoralized, frightened, worried, or lost. It is for those who are looking for an alternative means of healing physical illness or who wonder how to handle the pressure of each coming day."[27]

Many people ride just because they like it, and are receiving many of these benefits without even thinking of their horse as a therapist.

## Pets as counselors

People often talk to their animals, and some confide in them on a regular basis. This can often be of great help. It is as if the animal acts as a counselor. A woman in Chicago wrote to me about her Bernese Mountain Dog: "When I was sad he came and nudged me as if he wanted to say, 'Don't forget I'm still around!' When he lay down and I told him my troubles he looked at me understandingly with his big eyes and suddenly put his paw on my hand. From then on he has been doing that regularly."

Dr. Mary Stewart, of Glasgow University Veterinary School, was a leading researcher on human-animal interactions and also an experienced counselor. Her familiarity with both these areas has enabled her to compare pets, especially dogs, with counselors. It is generally agreed that a good counselor is "genuine, honest, empathic, nonjudgmental, able to listen, not talking too much, and ensuring total confidentiality." Mary Stewart points out that these are the very qualities that owners of dogs and other companion animals say they value so highly. It is as if these animal companions are quietly providing a counseling service to their owners without anyone else being aware of it. She suggests that some dogs and other animals increase their owners' self-esteem and feeling of well-being because they offer "congruence, empathy, and unconditional positive regard, the necessary conditions of any counselor who endeavors to provide a 'growth-producing climate' in which clients may get in touch with their own inner resources for development."[28]

Of course there are major differences. The very fact that animals live so much in the present and are unable to speak means that they cannot help their owners explore the past, look at personal relationships, and examine self-destructive patterns that keep repeating. Here good human therapists are irreplaceable.

Animals do have advantages, though. Humans, like other primates, find physical contact comforting. Especially when they are young they need to be touched and held lovingly to feel secure. And animals can comfort us by touching us, and we can stroke or cuddle them. A counselor, of course, has to be careful about offering physical reassurance, to avoid possible accusations of abuse.[29]

Perhaps the greatest gift animals can offer is their capacity for love. For clients with low self-esteem, it is difficult to accept that any human can have much regard for them, and so it is hard to feel that counselors really accept them, rather than just seeming to do so. Some fear that if all were revealed, the acceptance would be withdrawn. By contrast, they can easily believe that their animals love them unconditionally. And as Jeffrey Masson shows so vividly in his book of that title, "dogs never lie about love."[30]

## Animals that know when people are about to die

Some hospitals and hospices have resident pets, and in several cases these animals show a surprising ability to know when people are about to die and then stay with them until they do so. For example, Ginger, a caramel-colored part-Chow, lived in the Hubbard Hospice House, in Charleston, South Carolina. She turned up as a stray in 2001 and was much loved by patients and by their visitors. When one of the patients was about to die, Ginger lay under the bed until the person had passed away.[31]

The most famous death-anticipating animal in recent years is Oscar, a gray and white cat, who lives in the dementia unit at the Steere House Nursing Center in Providence, Rhode Island. He is not particularly friendly and is often aloof. But he seems to detect when people have less than about four hours to live, sitting beside them until they die, often purring and gently nuzzling them. He is more accurate than the doctors. For example, on his thirteenth correct call, a doctor thought one patient was about to die: She was breathing with difficulty and her legs had a bluish tinge. Oscar did not stay in the room with her, so the doctor thought the cat had finally gotten it wrong. But to the doctor's surprise the patient lived for another ten hours; Oscar returned to join the woman for her last two hours.

Oscar made medical history in 2007 by being written up in the prestigious *New England Journal of Medicine* by Dr. David Dosa, who summarized his achievements as follows: "Since he was adopted by staff members as a kitten, Oscar the Cat has had an uncanny ability to predict when residents are about to die. Thus far he has presided over the deaths of more than twenty-five residents. . . . His mere presence by the bedside is viewed by physicians and nursing home staff as an almost absolute indicator of impending death, allowing staff members to adequately notify families."[32]

## Dogs faithful after death

The devotion of some dogs continues after their person has died. Sometimes their devotion is so striking that they unwittingly not only achieve fame and a place in popular folklore but have monuments erected to them as well. There is one by the lonely waters of the Derwent Dam in Derbyshire, erected by public subscription and inscribed as follows:

*In commemoration of the*
*devotion of*
*Tip*
*the sheepdog who stayed*
*by the body of her dead*
*master, Mr. Joseph Tagg,*
*on the Howden Moor for*
*fifteen weeks*
*from 12th December 1953*
*to 27th March 1954*

Tip's master was a retired gamekeeper, aged eighty-one, who was found dead on the high moors fifteen weeks after setting off with Tip from his home in Bamford for a ramble over the hills. Search parties failed to discover them, for snow had covered the hills, and they had long been presumed dead. Three and a half months later a couple of shepherds came across the body of Joseph Tagg with Tip beside it, in a piteous condition but still alive. She rapidly became a national heroine and spent her final year in luxury at the home of her master's niece, who had to protect the dog from hosts of admiring visitors. A vast crowd assembled for the unveiling of her memorial, and pilgrims still visit her shrine.[33]

A similar celebrity attended the terrier belonging to a young man called Charles Gough, who died in a remote part of the Lake District in 1805. His remains were found months later by a shepherd, attracted to the spot by the emaciated dog still hovering around the corpse. Sir Edwin Landseer immortalized him in a painting, and a host of poets and artists added their own tributes.[34] The greatest of them, William Wordsworth, commemorated the dog in his poem "Fidelity," which ends with these lines:

Yes, proof was plain that, since the day
When this ill-fated traveler died,
The dog had watched about the spot,
Or by his master's side:
How nourished here through such long time
He knows, who gave that love sublime;

And gave that strength of feeling, great
Above all human estimate!

Countless other dogs never achieve such fame yet show deep devotion to their people after death. They are often grief-stricken and go through what can only be described as a period of mourning. Some lose the will to live and pine away. For example: "Immediately after the death the dog refused all food and itself died about a fortnight later," said one observer, Joan Creighton. A few bereaved dogs even seem to commit suicide by jumping out of windows or running out under lorries.

Some somehow find their owners' graves and stay there, like Greyfriars Bobby, the famous faithful dog of Edinburgh. Others visit the grave regularly but still come home. Molly Parfett of Wadebridge, Cornwall, wrote of such a dog: "My husband suffered a severe stroke in 1988 and died in hospital after having been there for two weeks. After his burial in a churchyard near our house, Joe, his dog, would disappear for hours, and we discovered that he was sitting by my husband's grave. How did he know when my husband died, and where he was buried?"

Such stories of enduring devotion illustrate how strong the bonds between dogs and their owners can be, and reinforce their age-old reputation for loyalty.

Chapter 6

# Distant Deaths and Accidents

If invisible bonds between animals and owners enable them to respond to each other's needs and also enable some pets to know telepathically when their owners are heading home, it would be surprising if these bonds were not affected by the distress or death of the owner.

The effects of death and distress are not subjects that lend themselves to experimental investigation. Obviously, no one can be asked to have an accident for the sake of science, or to die at a randomly selected time so that the reactions of their pet can be observed. The evidence necessarily comes exclusively from spontaneous cases.

On our database there are currently 177 accounts of dogs apparently responding to the distant accident or death of a human companion, 62 accounts of cats doing so, and 32 of humans knowing when their pet was in distress or had died at a distance. What can we learn from these cases?

## Dogs and distant accidents

Sometimes dogs show unmistakable signs of distress for which no immediate reason can be found. It later turns out that their owner was at that very time in danger or had an accident. Hilde Albrecht of Limbach, Germany, reported one such case: "One day our dog acted up like mad, jumped at the door, and wanted to get out. We locked her in, but she continued howling, scratching, was not herself. Suddenly my husband came home. He was injured because there had been a fight in the bar. The dog had known it. We do not know how."

In a case such as this it is scarcely conceivable that the dog could have known about the injury by sight, smell, or hearing. Nevertheless, skeptics might argue that the bar must have been close enough for the dog to have sensed some clue. But often the accidents occur many miles away from home, beyond all the range of all known senses.

One summer evening in 1991, for example, a young British soldier left his home in Liverpool to return by train to his barracks in southern England. Later that evening the family dog, Tara, started whining and shivering violently. The boy's parents thought she must be ill, gave her some acetaminophen and tried to comfort her. But she would not calm down for over an hour. She remained alert and restless until the telephone rang. According to Margaret Sweeney, "The phone call was from a Birmingham hospital to say that David had fallen from the train in the Tamworth area [80 miles away]. His injuries, though severe, were not serious, and they allowed him to speak to us. Tara showed her delight during the phone call, then lay down and went to sleep. We learned afterward that she first got upset at the moment he fell off the train, and calmed down when he was in hospital having been examined and made more comfortable."

There are forty-eight cases on the database of dogs reacting to distant emergencies in a comparable way, by showing signs of distress or restlessness. As well as the two examples given above, ten involve car or motorbike accidents, three falls, two accidents at work, one a capsized kayak, one a fire, and one a heart attack. Another took place when a woman was giving birth in a hospital 16 miles away.

I have personally experienced a dog's reactions coinciding with a distant accident. During the school half-term holidays in February 1998, we were looking after a yellow Labrador named Ruggles, who belonged to our friends and neighbors, the Beyer family. The son, Timothy, was away on a school skiing trip in the Italian Alps; his parents had gone on holiday in Spain. Ruggles settled in well and spent most of his time in our family room. But one morning when he returned from a walk at 11:30, he would not leave the entrance hall. All persuasion failed. He remained by the front door until he was taken out for another walk at 3:00. So striking and unusual was his behavior that I thought that Timothy's mother and father must have decided to come home early. I was expecting a telephone call from them to say that they had just arrived.

There was indeed a telephone call that afternoon, but it was not from Timothy's parents. It was from Italy, to say that Timothy had fallen off a chairlift that morning and broken a leg; he had been flown by helicopter to a hospital. The accident had happened at 11 A.M. British time. Curiously enough, when Ruggles returned from his afternoon walk, he was limping. He had jumped onto some broken glass and had a bleeding paw and a severed tendon. He had to spend the night at a veterinary clinic. So he and Timothy were both in the hospital at the same time with bandaged legs.

Obviously it is impossible to know for sure whether the dog's reactions between 11:30 A.M. and 3:00 P.M. were really due to the boy's accident. Ruggles did not seem particularly distressed when he was waiting by the door. Rather, it was as if he knew something important was happening and felt he had to be ready. But his reaction was so definite, and the coincidence so remarkable, that I think there could well have been a causal connection.

In these cases the dogs' reactions were of no help to the injured person. Apart from anything else, the animals were too far away. But in some cases dogs have helped save their owner's life, or tried to do so.

In one, the dog's owner had fallen out of a kayak in the middle of the Rhine River and had to struggle to save himself. "In my weak condition I saw my friends running toward me with my dog," the owner reported. "She was pulling them with her and barking loudly. They asked whether I had any trouble because the dog had suddenly started to tear at the lead and wanted to go down to the river—at exactly the moment when I had almost given up in my struggle against the water."

In Northern Ireland, a German Shepherd Dog, Chrissie, saved the life of his owner, Walter Berry, who got soaked with gasoline while repairing a car and then accidentally set himself afire with a welding tool. The dog was 200 yards away, through a couple of double garages and a yard, with Walter's wife, Joan, when this happened. "Chrissie went berserk and made noises that she had never made before," Joan said. She realized something was wrong and let Chrissie out. He rushed straight toward

Walter. Joan followed and fortunately arrived in time to put the fire out. Chrissie had saved Walter's life.

In these two cases, precisely because the dogs were close enough to be of help, it is hard to rule out the possibility that they were alerted by sound or other sensory clues. However, this objection cannot apply to a dog in San Francisco named Lupé, who saved her owner's life when she was over 40 miles away: Leone Katafiasz told this story: "When Lupé was about two years old, I had taken an overdose of drugs on a day when she was visiting with friends in San Jose. It was reported to me afterward that Lupé had suddenly gone to the end of the property and begun to howl 'uncannily,' and her agitation could not be relieved. After some time, my friends thought, 'Something must be wrong with Leone,' and they rushed to San Francisco and found me."

In many cases where dogs howl for no apparent reason, or show other obvious signs of distress, it later turns out that their owner was not just in danger but actually dying. Nothing that the dog could do would have saved them.

## Dogs that howl when their owners die

Out of 129 accounts I have received about the reaction of dogs to the death of an absent person to whom they were attached, 71, or 55 percent, involved vocal responses. Forty-seven of the dogs howled, five whimpered or whined, seven barked in an unusual way, eight cried, and four growled. In the cases where no sounds were mentioned, the dogs were said to be upset, miserable, shivering, terrified, or distressed.

The most impressive cases are those in which the animal shows clear signs of distress at unexpected times, especially when the person and the animal are far apart. In the following example, reported by Iris Hall of Cowley, Oxford, dog and owner were separated by more than 6,000 miles during the Falklands War:

> My son was very close to our West Highland White Terrier. He joined the Royal Navy in 1978 and, being shore-based much of his time until 1982, was home regularly for weekends. He traveled by train. We gradually came to realize that the dog would start getting excited

twenty to thirty minutes before he walked in the door, so as soon as she started running backward and forward to the front door, I would start getting him a high tea so that when he walked in, always hungry, his meal was ready. We used to laugh about it at the time. In April 1982 his ship, HMS *Coventry*, was drafted to the Falklands. Early evening May 25 the dog leaped onto my knee shivering and whimpering. When my husband came in, I said, "I don't know what's wrong with her, she's been like this for more than half an hour. She won't be put down off my knee." On the nine-o'clock news, it was said that a Type 42 had been sunk. We knew it was HMS *Coventry*, although the name wasn't given out until the next day. Our son was one of those lost. Our little dog pined away and died in a few months.

Typically, the dog's distress or its howling can be understood only in retrospect. Two such cases were reported by Stephen Hyde of Acton, London, and Mrs. G. Moore of St. Albans, Hertfordshire:

My brother Michael was a copilot in a Wellington bomber during the war. He went on many raids over Germany in 1940. At that time we had a dog, Milo, who was half spaniel, half Collie, and was particularly fond of Michael. One night in June, Michael was on his way home from a raid when he radioed to base to say that he was just off the coast of Belgium and would soon be back. That same night Milo, who slept in a stable at the back of the house, howled so much that my mother had to get up and bring him into the house. Michael never returned from his mission that night. He was reported missing, believed killed, June 10, 1940.

My husband and I were on holiday in County Cork in Eire in April 1968, and on Easter Saturday my husband died very suddenly. Our seven-year-old Standard Poodle was staying with friends in St. Albans. At just after midnight the poodle howled and rushed upstairs to my friend, who was in the bath. At just after midnight my husband died.

If the bond between person and animal is indeed a real connection, linking them together invisibly even over thousands of miles, then the disruption of this bond through the death of one, or through severe danger, might be *expected* to affect the other. To take a simple analogy, if two people are connected by a stretched elastic band, and one of them shakes it or lets it go, the other feels a difference. Even if one does not know exactly what is happening to the other person, he or she does know that *something* is happening.

It seems very unlikely that dogs form such bonds only with people. They are social animals and can form strong connections with each other. Do dogs react when other dogs to whom they are attached die in distant places? Sometimes they do. Here is an example sent in by Dr. Max Rallon of Châteauneuf le Rouge, France. This is just one out of thirty-two cases in our database in which the death of the other dog took place unexpectedly and at a distance:

> I have a Beance Sheepdog, Yssa, two years old, who came with me to France at the age of three months from the island of Réunion in the Indian Ocean, 10,000 kilometers away. There I left her mother, Zoubida, aged ten. On February 13 this year, Yssa was sleeping in my son's room. About 3:00 A.M. she came scratching at my door, whining, crying, and excited. She didn't want to go outside. At 9:00 A.M. my brother-in-law called from Réunion. The guard of our house had found Zoubida dead. She had been poisoned.

The existence of so many independent accounts of this type persuades me that this is a real phenomenon, even though it is not possible to do experiments. But further research is needed through the collection of more well-documented stories, the most convincing being those involving several witnesses to the dogs' behavior.

## Why do dogs howl when their owners die?

Howling is not found in all species of the dog family. Foxes do not howl, for instance, but only members of highly social species, such as domestic dogs, dingoes, coyotes, and wolves.[1]

The literature on wolf ecology suggests that wolves howl for two main reasons: first and foremost, to help assemble the pack, particularly before a hunt, and second, to seek contact with other pack members or to attract other wolves during the breeding season.[2]

Some wolves and dogs howl at the moon or the sky; no one knows why. And some howl in response to the sound of someone singing or playing the violin, as if they are trying to sing along. But like lone wolves, domestic dogs most often howl when they are alone, deprived of the company of humans or other dogs, especially if they are shut away. Desmond Morris says this "howl of loneliness" is a way of saying, "Join me."[3] So how can we explain their howls when a close companion dies?

Some of the dogs that howled when their owners died were shut outdoors, and their howling caused them to be brought inside, as in the case of Milo. In this limited sense their howling worked and brought them companionship and comfort. But in many cases the dogs howled when they were not shut out, and the attempts of people to comfort them were unsuccessful, at least to start with. Perhaps this kind of howl is a way of expressing grief. And it may have a long evolutionary ancestry, because several observers of wolves have found that "wolves howl in a particularly mournful way when a beloved companion has died."[4]

The animals that did *not* howl were clearly very disturbed or upset. They obviously felt something was wrong. Perhaps they did not know what; they may simply have been apprehensive or fearful. If human companions were nearby, they usually went to them for comfort.

## The responses of cats to distant accidents and deaths

Although fewer cats than dogs seem to react to accidents and emergencies, the situations in which they do so are similar, as the following example, sent by Andrea Metzger of Bempfingen, Germany, shows:

> In May 1994 I sat outside on the veranda, and our three-year-old Persian cat, Klaerchen, lay beside me, purring comfortably. My eleven-year-old daughter had gone out with her girlfriend on her bicycle. Everything seemed wonderful and harmonious, but suddenly Klaerchen jumped up, uttered a cry that we had never heard before and in a flash ran into the living room, where she sat down

in front of the shelves with the telephone. The phone soon rang and I got the news that my daughter had had a bad accident with the bike and had been taken to hospital.

Cats most commonly respond to distant deaths by making unusual sounds, such as howls or plaintive meows, or they may whine and show other signs of distress. Hedwig Ritter of Zurich, Switzerland, wrote of one such cat:

> We had a beautiful Carthusian tomcat that we all loved, but he loved my husband most of all. In the summer holidays we went camping in Denmark and left the cat at an animal home in Switzerland. In Denmark my husband, who was forty-eight years old and had never been ill, died of a heart attack. When we went to pick up our cat, the lady told us she knew exactly when a tragedy had happened to us and then gave us the exact day and hour, which she could not have known! Our tomcat had withdrawn into a corner and whined in a way he had never done before, staring at a certain point in front of him as if he observed something special, his whole body shaking.

But while most cats seem to respond vocally to the death of a distant person, some react silently. One cat simply hid on the night the father of a family was dying in the hospital: "Nobody could find him," wrote Madame Charlin of Lyons, France. "He never went out. He only came out of hiding when we came back for the burial." Other cats changed their sleeping places.

The realization of the person's death does not seem to diminish with distance. In some of the cases in our collection the person who died was thousands of miles away, yet the cat still seemed to know. For example, a tomcat belonging to a family in Switzerland was very attached to their son Frank, who went away to work as a ship's cook. He came home irregularly, and the cat used to wait for him at the door before he arrived. But one day the cat sat at the door and meowed in extreme sadness. "We could not get him away from the door," wrote Karl Pulfer of Koppingen, Switzerland. "Finally we let him into Frank's room, where he sniffed at everything but still continued his wailing. Two days after the cat's strange behavior [had begun] we were informed that our son had died at exactly that time on his voyage, in Thailand."

## Human reactions to distant deaths of animals

If, as I have suggested, the bond between animal and person can be thought of as being like an elastic band, then it should allow influences to pass in both directions, from person to animal and from animal to person. We have already considered influences passing from a person to an animal. What about influences in the other direction? Do some people react to their animals when they have had accidents or are dying at a distance?

Judging from the numbers of reports in the database, humans are generally less sensitive to their animals than animals are to their people. We have 239 reports of animals reacting to the distant death or distress of people and only 32 the other way around, of which 26 came from women and 6 from men. Most concerned dogs and cats. Most took place when the people were awake, but some occurred in dreams.

The waking experiences typically involved feelings of worry and distress, and some involved physical symptoms as well. For example, on May 20, 1997, Diane Arcangel of Pasadena, Texas, was leaving a hotel to go to the airport to catch a plane home. Soon after the car journey began, at 4:05 P.M. Texas time, she began to feel agitated, but could find no reason for it. She wrote as follows:

> As we continued the drive, I began to feel nauseated and to perspire. After about fifteen minutes I was feeling that my stomach and intestines were being torn so intensely that I held my stomach and bent over. By the time we arrived at the airport, I felt physically sick and in deep grief. Fearful that something was very, very wrong at home I called my daughter. "We just had a terrible storm with lightning, but it is over now," she said, but told me everything was fine. But I cried all the way home. When I arrived at Houston airport at 10:00 P.M. I found my husband in tears. He explained that lightning had hit our house at 4:08 P.M. (All our clocks were stopped at this time.) Kitty, one of my eight cats, was so terrified of the storm that she ran outside. When my

husband got home, he saw two large dogs in the back yard, standing over her lifeless body. As he pulled them away, he could see they were both covered with her blood and hair. The trauma to her body was where I felt excruciating pain at the same time it was occurring to her.

Some other women have also experienced physical distress, but in a less specific way. In the case of Mary Wall, who lives in Wiltshire, England, it occurred when she was more than 2,000 miles away from her Shih Tzu dogs, on vacation with her husband in Cyprus: "At 4:00 P.M. Cyprus time on a Friday I was overcome by an intense sensation, so much so that I eventually mentioned it to my husband. Something was desperately wrong with the dogs. The sensation was so strong it was physically distressing. On arriving at Heathrow airport a few days later I was told the boy dog died the previous Saturday. I would never have believed a dog or any animal could 'get through' to a human, though I have experienced 'knowing' what was happening two or three times in my life, but only concerning close relations."

Some correspondents felt that something was wrong but the feeling was not specifically connected with the dog. For example, Lotti Rieder-Kunz, a Swiss woman working in her office in Basel, had a strange feeling one morning. She mentioned it to her co-workers but could not explain it. "After about one hour the thought went through my head, 'You should ring up home.' I learned that an hour before our [German Shepherd Dog] had been hit by a car and died."

In other cases, the knowledge that the dog had died was quite explicit. Nancy Millian, of New Haven, Connecticut, had gone on vacation leaving her dog, Blaze, at home. "About five days into the trip, I became incredibly agitated and heard the words, 'Blaze died' in my head. I told my friend, who responded that I was probably just experiencing normal worry. I called home and was assured that all was well." Two days later she arrived home to learn that her dog had indeed died the day she felt agitated. His caretaker had not wanted to upset her, knowing there was nothing to be accomplished by her early return.

Keith Phillips lived in Australia about 15 miles from his ex-wife, who had kept their dog, Blacky, when they divorced. He had not seen Blacky for months. "I awoke one morning in March 1996 in a panic," Keith wrote. "I had had a most distressing, vivid dream concerning Blacky. I felt most strongly that he was very ill and that he was calling for me. I did not have my ex-wife's phone number, but midmorning she called me. She explained that Blacky was very ill, and would I like to see him. I immediately left work to see him. He was so ill that he could hardly move, but even so, he managed to lick my hands." The dog died the next day.

Finally, here is an example of explicit information coming through in the dream of a teenage girl named Laura Broese: "In the summer of 1992 I was away from home the month of July. One night I had a nightmare that my cat had been run over by a car in our road (we were living in Belgium at the time, and I was in Holland). I remembered the dream the next morning, and since I kept a diary at the time, I wrote down my dream. When I got home I was told that my cat had been run over. I checked my diary and it was the same night that I had had my dream."

In all these cases, the people and their animals were far apart and there could have been no transfer of information through normal sensory channels. Telepathy, or something like it, seems to me the only plausible explanation. Just as animals can react telepathically when their owners are in distress or dying, people can be influenced in a similar way by their animals' distress or death.

## People who know when other people have died

The phenomena described in this chapter concern people and nonhuman animals. Similar reactions to distant deaths and accidents can occur between people. Indeed, some of the most impressive cases of human telepathy concern people at a distance who are in danger or who are dying. Through surveys and questionnaires, the pioneers of psychic research built up impressive collections of such cases, authenticated through careful inquiry and attested to by witnesses.[5] In just over half these cases, people dreamed of the person who was dying or in distress. Of those cases that occurred when people were awake, the majority involved an impression or intuition without a visual image. About 20 percent of the total number of cases involved images or hallucinations.[6]

Of the cases in the database in which people seemed to know about the distress or death of distant pets, one involved visual imagery—the dream about the cat being run over. The others involved feelings, impressions, or intuitions. So the same kinds of experience seem to occur with pets as with other people, even though the proportions of dreaming and waking, or visual and nonvisual communication, may differ. Just as most cases of person-to-person telepathy depend on close relationships, so do the person-to-pet and pet-to-person cases discussed in this chapter.

Part IV

# Intentions, Calls, and Telepathy

*Chapter 7*

# Picking Up Intentions

People's intentions, calls, and commands can affect their companion animals, and animals' needs and emotions can likewise affect people. In some cases, these intentions, calls, and needs seem to be communicated telepathically. This telepathic ability exists within animal societies in the wild, demonstrating that telepathy has a long evolutionary ancestry. Telepathy is natural, not supernatural, normal, not paranormal, and is an important aspect of animal communication. Human telepathy needs to be seen in this broader biological context.

I start here by considering the ways in which animals pick up their owners' intentions, and people pick up their pets'.

## Animals "reading people's minds"

Many people have noticed that their animals seem to read their minds. The perceptiveness of pets may well depend on a combination of influences, such as close observation of body language, hearing particular words and tones of voice, and learning the owners' routines. In addition, they may be able to pick up intentions directly by a kind of resonance or telepathy. As we have seen, some animals pick up people's intentions and feelings when they are miles away, so it would be surprising if they could not do so when close by.

Many people who are experienced with animals take telepathy for granted, and a wealth of anecdotal material points to the reality of telepathic influences. Committed skeptics, of course, believe that any mysterious connections currently unknown to science are either impossible or too unlikely to merit serious attention.

The only way to resolve this issue is to take a close look at the evidence from people's experiences with their pets, and then to carry out experiments to clarify what is going on.

## Cats that disappear before visits to the vet

Some cats strongly dislike going to the vet. Dozens of cat owners have told me that their cats simply vanish when they are due to be taken for their appointment. For example, "We had a cat who could read minds," Eugenia Potter said. "The most outstanding example was a time when she was cuddled up on my chest early in the morning. I was thinking about what I needed to do that day. When I reached the point when I thought, 'and I have to take Patches to the vet,' she leaped up, ran away, and hid. I had to cancel the appointment because I couldn't find her."

Experienced owners of such cats try to avoid giving away any clues, but often their efforts are in vain. According to Andrea Künzli of Starrkirch, Switzerland, "The cat always knows hours ahead of time when I am going to take him to the vet, long before I actually fetch his basket from the attic. I try to act as natural as possible so he won't notice, but he can see through me at any time and will yowl to go out." This is inconvenient not only for the owners but also for the vets. Some advise people to keep their cat shut up indoors before the appointment, especially when injections or operations are involved. But some cats still escape.

How common is this kind of behavior? We carried out a survey of the veterinary clinics listed in the North London Yellow Pages. We interviewed the vets or their nurses or receptionists, asking whether they found that some cat owners canceled appointments because the cat had disappeared. Sixty-four out of sixty-five clinics had cancellations of this kind quite frequently. The remaining clinic had abandoned an appointment system for cats: people simply had to turn up with their cat, and thus the problem of missed appointments had been resolved.

Although there was general agreement that some cats do indeed pick up their owners' intentions, there was a variety of opinions as to how they might do it. A veterinary receptionist in East Barnet said, "It is not always the cat basket. The clients know that once they produce the basket there is not a hope in hell of catching the cats, so it is usually before the baskets have been brought out. People say they get home around 5:30 P.M. and the cat is always on the doorstep, but the day of the appointment he is not there. I think they have definitely read their thoughts because the owner has not been in all day so they cannot have seen that the owner is upset or behaving any differently. They say, 'I don't know why he hasn't come back for his tea. It is very odd.' "

A veterinary nurse in Wembley reported that "Sometimes people say they have gone to get the basket out and the cat has stayed in the bushes in the garden, or cats don't always come back in the morning and that's before they have seen the basket. They are expected to come in for their breakfast and they stay in a tree. Or for evening appointments, the people go out to work and in the meantime the cat goes and hides."

## Other feline aversions

Visits to the vet are not the only thing cats try to avoid. Some also run away when they are going to be given medicine, sprayed for fleas, or subjected to other procedures they dislike. Sheila Howard of Wandsworth, London, had this to say:

> My cat, Ciggy, knows where a lot of his food comes from and often takes up a position nearby waiting expectantly for his next meal. When, however, I go to the same cupboard to get his spray to treat his coat, even before I get hold of the spray he makes a dash to go out through his cat flap into the garden to avoid being sprayed. I never tell him when I am going to spray him and I have even tried thinking of something else while going for the spray, but he always seems to sense my intention.

Cats also tend to disappear before they are going to be taken away for good. Pauline Westcott of Roehampton, Surrey, for many years did rescue work with cats, collecting them in response to telephone calls from their finders or from people who no longer wanted them. In nine out of ten cases the cats had to be destroyed:

> We continually found that if an appointment was made to collect them, even with strenuous efforts on the part of the keeper, the animals would in many cases not be found at all. We were told the cat disappeared within minutes of the call being made to make an appointment, or even just before the call was made. Only by boarding up a room and literally shutting every single entrance, crack, ventilator, etc., could we be sure of finding the animal. So much time, gasoline, and man-hours were wasted that our system had to be altered time and time again. There were inevitably visits where a cat was never captured.

The cats' awareness of imminent danger was obviously of survival value, and if wild animals have comparable abilities, they would presumably be favored by natural selection. But even less is known about such intuitions in wild animals than in pets.

Compared with cats, dogs rarely disappear or try to hide before going to the vet. However, some seem to know when they are on the way to the clinic. Maxine Finn, a veterinary receptionist in North London, described their reactions: "A lot of . . . dogs know when they are coming to the vet. When they are driving the dog starts shaking and whining as if they know they are on their way. About one client a week says that. Sometimes we have clients that come back a few years after their last visit and the dogs still start to shake on the way. They either seem to remember the route or they somehow pick up where they are going to."

Some dogs, like cats, anticipate when they are about to be subjected to procedures they dislike, such as being washed or having their nails or fur clipped. Sylvia Scott of Goostrey, Cheshire, England, reported this reaction: "On the day our poodle, Snowy, was going to be clipped (about every six weeks), no matter what precautions we took to stop her knowing, she always crawled under the piano or a bed to hide. To this day I shall never know how she knew, other than by reading my mind."

## Dogs that anticipate going for a walk

The commonest way in which dogs respond to their owners' intentions is not through aversion but through their enthusiasm for walks. Most dogs are excited by the prospect of a walk and react with eager anticipation when they see their owners preparing to take them out or when they hear words such as "walk." Some dogs are walked routinely at the same time every day, and they get excited as that time draws near. If they are usually taken out after a particular TV show has ended, for example, they show signs of anticipation when they hear the end-title music or see the set being switched off. In this respect their reactions are like the conditioned reflexes studied by the Russian physiologist I. P. Pavlov, who found that when dogs were repeatedly given meat after they heard a bell ring, they came to associate the bell with feeding and salivated when it rang, before they even saw the meat.

But many dog owners do not take their dogs out at routine times, and some have found that their dog's excitement begins *before* they have given any obvious signs, such as putting on their coat or getting out the lead. For example, Douglas Walker of Lansing, Michigan, found this happened with his dog, Patrick. "Several times I was at home, just Patrick and me, watching TV, and he was asleep in front of the TV. Quietly I would think about when I was going to take him out for a walk, and no sooner had I started thinking about walking him, and *zap*! Up he wakes and comes over to where I am sitting and starts getting excited, almost as if I said out loud, 'Do you want to go for a walk?' I had said nothing and had not moved in any way, but my thoughts seemed to get through to him."

Many other people have noticed that their dogs seem to pick up their intention to go for a walk, even when they are not giving any visible or audible sign. We have more than 170 such reports on our database. Most informants are well aware of the possibility that body language could give the game away, but some have come to the conclusion that this cannot always be the explanation, because their dogs still react when sleeping or out of sight. For example, Sue Stickley says, "My dog knows if I'm thinking of going for a walk even when I'm in a different room.

I have experimented with her over the years and have noticed that no matter what time of day it is, if I visualize us going out for a walk she immediately comes running and gets excited."

Mary Rothwell of Arnold, Nottingham, says, "I could be doing anything or nothing, just sitting sewing or baking, and a thought would come in my head: 'Go for a walk. Take the dogs. It's nice out,' and the Dachshunds would be there at my feet wagging their tails. They could not tell from my expression or movements, as this happened when they were in the garden or fast asleep. I deliberately tested this theory, and no way could they see me. Once the thought was in my head, my dogs knew, no matter what they were doing at the time."

With dogs that respond in this way, it is possible to do simple experiments in which the dogs are kept where they cannot hear, see, or smell their owner, and they are videotaped continuously. Then at a randomly selected time the owner starts thinking of taking them for a walk, and after a delay of, say, five minutes, actually does so. Does the videotape show that the dogs display signs of excitement before going for their walk and after their owner formed the intention of taking them?

Some preliminary experiments of this kind have already been carried out at my request by Jan Fennell of Winterton, Lincolnshire, England. She is an animal behaviorist, well aware that animals can pick up routine patterns or clues from their owners' behavior. She has six dogs, and had already observed that the dogs seemed to know when she intended to take them out at nonroutine times, even when she tried to avoid giving them any clues.

For the purpose of the experiment, the dogs were shut up in an outbuilding where they were taped continuously by a video camera on a tripod, pointed toward the door. She thought about taking them for walks at randomly selected times while the camera was running. She carried out the experiments on five different days: once in the morning, twice in the afternoon, and twice in the evening.

The videotapes show that most of the time the dogs lay around or played together. Now and then some of the dogs reacted briefly, with pricked ears to sounds from the outside, such as passing motorbikes. But after Jan had decided to take them for a walk, in four out of the five videotapes, the dogs moved closer to the door and sat or stood in a semicircle around the door, some with their tails wagging. They remained in this state of obvious anticipation for three to five minutes, until Jan came and opened the door to let them out for their walk. By contrast, in the remaining videotape, they showed no such advance reaction, and responded only thirteen seconds before she entered the outbuilding, probably in response to the sound of her approach.

In addition, in a control experiment Jan shut the dogs up and visited the outbuilding at a randomly selected time but with no intention of taking them for a walk. When she went to the outbuilding, only one of the dogs went to the door a mere twelve seconds before she opened it. When she opened the door, the other dogs got up and moved around, but remained quite calm and showed none of the excitement that preceded a walk.

These pioneering experiments suggest that the dogs, on most but not all occasions, could indeed anticipate their owner's intention to take them out without being able to see her.

## Dogs that know when they are being taken out by car

Some dogs anticipate when they are going to set off with their owner in a car. This phenomenon is similar to the anticipation of walks. Although routines or normal sensory clues may often account for the dogs' reactions, this is not always the case. Here is an example from Dieter Eigner of Powelltown, Victoria, Australia:

> My wife and I leave home at irregular intervals to go shopping, etc. In general we take the dog along in our car. On departure the dog will sit next to the tailgate to be let into the rear compartment of the car. If I come out of the house with the intention to go away, the dog will dash to the tailgate. But if I leave the house with the car keys in my hand only to pick up something that I have left in the car, the dog will not respond. Today when we were at the table for morning tea, my wife said she would like to go shopping in about five minutes. Looking out of the kitchen window, I saw the dog already sitting next to the tailgate, facing the door of the house in expectation. At this point I had not left the house and had no prior contact with the dog. Absolutely no physical element, to our knowledge, could have indicated to the dog that we were about to go away.

On our database there are more than one hundred and twenty such examples, but there is no need to discuss them at length because of their general similarity to anticipation of walks by dogs.

## Pets that know when their owners are about to leave them

It makes a big difference in the lives of domestic animals when their owners go out, especially when they go away on vacation or on other protracted journeys. Many dogs and cats seem to pick up their owners' intention to leave. No doubt this often happens because the animals see obvious preparations such as the packing of suitcases. But on our database there are more than 180 reports from pet owners who think the animals know even before they have seen any telltale signs. For example, Mary Burdett of Blackrock, Ireland, wrote that "Our Labrador, if anyone was going on holiday, would go 'round looking miserable for three or four days before they actually left and before any packing started. Once they left he reverted to normal."

In the four surveys my colleagues and I carried out in Britain and the United States (page 48), we asked, "Would you agree or disagree that your pet knows you are going out before you show any physical sign of doing so?" This question covers several phenomena: going away on a journey, going out and leaving the animal behind, and going out and taking the animal along. On average, 67 percent of dog owners and 37 percent of cat owners agreed (Figure 7.1). These were the highest percentages of positive answers to any of the questions we asked about the perceptiveness of pets. But although this is one of the most common ways in which animals react to people's intentions, it is one of the hardest to test experimentally, because it is difficult to keep the person away from the pet for hours or even days before the time of departure.

## Animals that know when they are going to be fed

Many animals seem to know when they are going to be fed, often showing their anticipation through excited behavior. This expectation may often be a result of routine, of seeing, smelling, or hearing the person

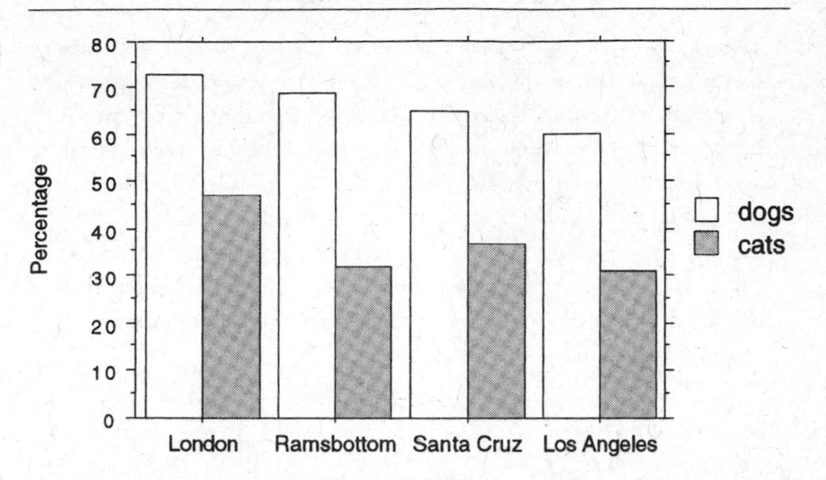


***Figure 7.1** The percentages of dog and cat owners who said their animals knew when they were going out before they showed any physical signs of doing so. The surveys were carried out with a random sample of households in London; Ramsbottom, England; and in Santa Cruz and Los Angeles, California.*

preparing the food, or of responding to some other signal from that person. But it may also depend on a more subtle detection of intent.

The most striking examples concern not regular meals but treats or tidbits. Said Frank Bramley of Telford, Shropshire: "The older my German Shepherd, Maxi, gets, the more he appears to be telepathic. I only have to think 'sausages' or 'chocolate' or 'biscuits,' and he appears. He can be in the garden, with the door closed and he *knows*. I can open the fridge a dozen times and no reaction, but then, if I take sausages or chocolate from the fridge, he is at the door and 'knocking' to come in." Likewise, Martha Markus of Vancouver, British Columbia, found that her dog, Wilkie, seemed to pick up her thoughts: "On many an occasion when she was outside sitting under our apple tree, I would consciously think about going to the fridge and getting her a piece of cheese. I would visualize doing this and then think to myself, 'Come in, girl, if you'd like some cheese.' Within three minutes, she would be scratching at the back door."

Cats show very similar behavior. Here, from Joan Hayward of Dorchester, England, is one example out of many:

Tiger, a tabby, used to weave in and out of my legs, picking up scraps of meat that fell to the floor—common enough—but she always seemed to *know* when I was just *thinking* of getting the food grinder out for meat (no reaction if I was going to grind fruit or vegetables), and would be under my feet, though previously curled up asleep or out in the garden—before I even opened the drawer in which the grinder was kept. This was also before I got out the meat (otherwise I would have assumed that she'd smelled it). She just seemed to read my mind, as she never reacted when I opened that same drawer to get out other items.

## Horses

Many horse owners and stablekeepers have found that their animals seem to anticipate being fed, but it is difficult to separate the direct effects of intention from routine and from hearing or seeing the food being prepared and fetched. However, some people who keep horses live miles from the stable or paddock, too far away for the horse to see, smell, or hear them, and in these circumstances, some horses still seem to anticipate their arrival, as if they know they are on the way, even if they come at nonroutine times.

Olwen Way of Brinkley, near Newmarket, England, used to keep a stud farm and racing stables and had many years of experience with horses. She later kept a pony named Freddy near her son's house in the next village, Burrough Green, a 2.5-mile drive from her home. Because he suffered from laminitis (an inflammation of the foot exacerbated by eating young grass), Freddy had to be kept in a separate paddock, and Olwen went to Burrough Green to feed him daily. Her daughter-in-law and grandchildren often noticed that Freddy went to the fence in his paddock and seemed to be waiting for Olwen before she actually arrived, even though she came at irregular times.

Over a period of six months, Olwen and her family kept a log of Freddy's waiting behavior, on the days when there was someone to observe him. Usually Freddy reacted two or three minutes before Olwen arrived in her car, but sometimes eight to ten minutes in advance, when she was setting off from home. On one occasion she came from a more distant village, a twenty-minute drive away, and Freddy reacted twenty minutes before she arrived.

We videotaped Freddy's behavior in experiments in which Olwen set off in response to telephone calls at randomly chosen times and traveled by taxi. Freddy still reacted in advance, ruling out the possibility that he was responding to routines or to sounds from Olwen's car.

## Bonobos

We have inquired about anticipation of feeding times by monkeys and apes at various zoos in Europe. In most cases the animals are fed at regular times, so it is hard to distinguish the effects of their keepers' intentions from routine. Also, like most zoo animals, they tend to be well fed, so they are rarely very hungry.

However, some apes do react in a way that suggests they do pick up their keepers' intention to feed them. For example, Jacqueline Ruys, a head keeper at Apenheul Zoo, Apeldoorn, Holland, looked after three bonobos (pygmy chimpanzees). She prepared their food in the early afternoon and kept it in a building 100 meters away from the cage, with trees and another building in between. The animals were usually fed between 3:00 and 5:00 P.M., but not at a fixed time. "When I am leaving our building with the food bucket in my hand," Ruys said, "they cannot see me but the males immediately start to scream. They start screaming when I have one foot out of the door. Yet when I go out to throw a bucket of trash in the can outside, without their food, they don't scream. I go in and out of the building where we prepare their food about fifty times a day. I don't know how they know, but they know when I am coming with the food instead of something else."

My favorite ape story is from Betty Walsh, senior chimpanzee keeper at Twycross Zoo in Warwickshire, England. It concerns her bonobos: "One bonobo had a long bamboo cane, which she was poking members of the public with, so we wanted it off her. I had a bag of four cakes, which we were going to have for our tea, and I thought I would give her a cake if she gave me the stick. But she saw I had four cakes and she broke the bamboo stick into four pieces, one piece for each cake. It was more than clever. She worked it out in a split second."

Here it is impossible to separate telepathy, subtle cues, and sheer intelligence. The ape somehow picked up her keeper's intention to reward her with a cake for giving up the stick, and having seen the four cakes immediately thought of a way of getting all of them.

## Parrots

Some language-using parrots respond to their people's moods, feelings, and intentions by making appropriate comments. In some cases their ability seems to be telepathic.

The fact that parrots can use language meaningfully has been established beyond reasonable doubt by Dr. Irene Pepperberg, now at Brandeis University, who spent nearly thirty years training and testing an African Gray parrot called Alex, who had a vocabulary of more than one hundred words. Through meticulous experiments, Pepperberg showed that Alex was capable of using words meaningfully and could grasp concepts like "present" and "absent" and use the words for colors appropriately, whatever the shape of the colored object he was shown. When Alex died in 2007, he had an obituary in the *New York Times.*

Before Pepperberg's research, most scientific studies of human-animal communication were with apes, using sign language. Pepperberg succeeded in showing that parrots rival apes in the ability to use thoughts and concepts, and have the huge advantage of being able to speak. She summarized her research with Alex and other parrots in a monumental book called *The Alex Studies: Cognitive and Communicative Abilities of Grey Parrots.*[1] As she remarked when Alex died, "He broke all of our preconceived notions about bird brains."[2]

Inspired by seeing Alex on television in 1997, Aimée Morgana, an artist who lived in New York State, began training a young male African Gray parrot, N'kisi (pronounced "in-key-see") in the use of language. She taught him as though he were a human child, starting when he was five months old. She used two techniques known as "sentence frames" and "cognitive mapping." In sentence frames she taught N'kisi words by repeating them in various sentences such as, "Want some water? Look,

I have some water." Cognitive mapping reinforced meanings that might not yet be fully understood. For example, if N'kisi said "water," Aimée would show him a glass of water.

By March 2010, when he was twelve years old, N'kisi's vocabulary had expanded to 1,500 words, the largest ever recorded for a parrot. He usually speaks in sentences.

Although Aimée's primary focus is on documenting N'kisi's use of language, she soon noticed that he said things that seemed to refer to her thoughts and intentions. He did the same with her husband, Hana. After reading about my research on telepathy in animals, she e-mailed me, summarizing some of her observations. Aimée wrote, "N'kisi regularly comments when we are thinking about eating, going out, or taking a shower, even if we are sitting quietly in another room and he sees no body language and hears no audio cues. At these times he will say, for example, 'You want some yummy?' 'You gotta go out; see ya later,' or 'You wanna take a shower?' "

For more than ten years, Aimée has kept a detailed log of N'kisi's comments and noticed a number of seemingly telepathic incidents. Here are some examples:

- "I was thinking of calling Rob and picked up the phone to do so, and N'kisi said, 'Hi, Rob,' as I had the phone in my hand and was moving toward the Rolodex to look up his number."
- "We were watching the end credits of a Jackie Chan movie, edited to a musical sound track. There was an image of him lying on his back on a girder way up on a tall skyscraper. It was scary due to the height, and N'kisi said, 'Don't fall down.' Then the movie cut to a commercial with a musical sound track, and as an image of a car appeared, N'kisi said, 'There's my car.' N'kisi's cage was at the other end of the room, behind the TV. He could not see the screen, and there were no sources of reflection."
- "I was in a room on a different floor, but I could hear N'kisi. I was looking at a deck of cards with individual pictures and stopped at an image of a purple car. I was thinking it was an amazing shade of purple. Upstairs he said at that instant, 'Oh, wow, look at the pretty purple.' "

N'kisi also seemed to respond to dreams. He usually slept by Aimée's bed. "I was dreaming that I was working with the audio tape deck. N'kisi, sleeping by my head, said out loud, 'You gotta push the button,' as I was doing exactly that in my dream. His speech woke me up." On another occasion, "I was on the couch napping, and I dreamed I was in the bathroom holding a brown medicine dropper. N'kisi woke me up by saying, 'See, that's a bottle.' "

I visited Aimée and N'kisi in April 2000 and was impressed by their remarkably close relationship and by the way that N'kisi used language meaningfully. After hearing about so many remarkable incidents, I was of course interested to see for myself if he could really pick up Aimée's thoughts. We did a simple test that replicated a situation in which N'kisi had appeared to demonstrate telepathy spontaneously. Aimée and I went to another room where N'kisi could not see what we were doing, and I watched as Aimée looked at several different images. When she concentrated on a picture of a girl, after several seconds N'kisi said with unmistakable clarity, "That's a girl." N'kisi was in a different room, and Aimée and I had not spoken about the image.

Clearly, it was important to test this apparent telepathic communication in controlled experiments that could be analyzed statistically. Aimée and I devised a procedure that was rigorously scientific and yet worked fairly naturally in N'kisi's familiar environment.

Aimée had noticed that N'kisi seemed to respond to moments of discovery. As she put it, N'kisi seemed to "surf the leading edge" of her consciousness. Therefore, methods that used repetitive images, like card guessing tests, were not likely to work. In order to preserve an element of surprise, we designed an experiment in which Aimée opened sealed envelopes one at a time, each of which contained a different photograph. A man who was not otherwise involved in this research selected the photographs in accordance with a list of key words in N'kisi's vocabulary. He sealed each photograph in a thick, opaque envelope, shuffled the envelopes, and numbered them in a random order. Neither Aimée nor I knew which pictures he had selected, nor what order they were in.

In each trial Aimée opened an envelope and looked at the picture for two minutes. Two synchronized cameras recorded Aimée and N'kisi, who were in separate rooms on different floors of the house. They could not see each other, nor could N'kisi hear Aimée. In any case, Aimée said nothing, as confirmed by the audio track recorded by her camera.

Subsequently, three separate people independently transcribed the tapes of N'kisi's comments. They did not know which pictures Aimée had been looking at. The transcripts were in good agreement with each other. They were then compared with the images Aimée was looking at in the synchronized videotapes.

In many cases, N'kisi's comments corresponded to the images Aimée was seeing. For example, when she was looking at a picture of flowers, he said, "That's a pic of flowers." When she was looking at a picture of someone talking on a mobile phone, he said, "What'cha doin' on the phone?" and made a series of noises like a phone being dialed. When she was looking at a picture of two people on a beach wearing skimpy swimsuits, he said, "Look at my pretty naked body."

N'kisi was right far more often than he would have been if he had simply been talking at random. In more technical terms, we considered it a "hit" when he said a predefined key word that corresponded to an image representing that key word. He scored 23 hits in a total of 71 trials.

The results were analyzed independently by a professor of statistics at the Free University of Amsterdam, Holland. His statistical analysis showed that N'kisi's hits were far more frequent than would have been expected by chance, and his errors far fewer. The results were highly significant statistically. Full technical and statistical details are given in our paper on this experiment in the *Journal of Scientific Exploration*.[3]

*Chapter 8*

# Telepathic Calls and Commands

In the preceding chapter, I discussed ways that animals respond to people's intentions. Many animals seem to pick up intentions whether their owners like it or not, without any deliberate effort on the owners' part.

In calling an animal or giving a command, an owner is deliberately trying to influence the animal's behavior. Someone who calls a cat wants it to come. A shepherd wants his dog to herd sheep in accordance with his intentions. A rider wants her horse to jump a hedge. Through calls and commands, people actively *will* their animals to do something. Sometimes these calls and commands appear to be communicated telepathically, and they can go in both directions, from people to animals and from animals to people.

Telepathy also seems to occur in connection with calls by telephone. Some cats and dogs seem to know when their owner is calling or is about to call. And before they actually answer the phone, many people have had seemingly telepathic intuitions that a particular person is calling.

## How common are telepathic experiences with animals?

Among those who work with dogs and horses, the existence of telepathic influences is usually taken for granted. "No one in their senses disputes them," said Barbara Woodhouse, the formidable British dog trainer:

> You should always bear in mind that the dog picks up your thoughts by an acute telepathic sense, and it is useless to be thinking one thing and saying another; you cannot fool a dog. If you wish to talk to your dog you must do so with your mind and willpower as well as your voice. I communicate my wishes by my voice, my mind, and by the love I have for animals. . . . A dog's mind is so quick at picking up thoughts that, as you think them, they enter the dog's mind simultaneously. I have great difficulty in this matter in giving the owners commands in class for the dog obeys my thoughts before my mouth has had time to give the owner the command.[1]

When I began asking pet owners, dog handlers, blind people with guide dogs, and horse riders about their communication with their animals, I soon found that Barbara Woodhouse's opinions on the subject are widely shared. This impression was confirmed by formal surveys.

My associates and I carried out surveys of randomly selected households in England and the United States in which we asked animal owners the following question: "Would you agree or disagree that your pet responds to your thoughts or silent commands?" An average of 48 percent of dog owners and 33 percent of cat owners agreed.

We then asked a further question: "Would you agree or disagree that your pet is sometimes telepathic with you?" The pattern of answers was broadly similar to that for the question about thoughts and silent commands. On average, 45 percent of dog owners and 32 percent of cat owners believed that their animals' responses were not simply a matter of picking up sensory clues but involved a telepathic influence.

We also asked people about their experience with previous pets: "Would you agree or disagree that any of the pets you have known in the past were telepathic?" In response, 45 percent of pet owners and 35 percent of people currently without pets said they agreed.[2]

These survey results suggest that at least one-third of all adults believe they have had or still have a telepathic connection with animals. In Britain, that would mean more than 15 million people, and in the United States more than 70 million!

What kinds of experiences have led so many people to think that their animals can respond to them telepathically? I have already discussed the ways in which animals seem to react telepathically to people's intentions and distress. I now turn to various kinds of calls and commands.

## Summoning cats

Of all the seemingly telepathic phenomena described by cat owners, the ability to summon a cat mentally is one of the commonest. For example, Nancy Arnold, of Kalamazoo, Michigan, has five cats. She says that when they are outside, "I only have to think about a certain cat, and within a minute or so, the cat appears at the door. I just take their telepathy for granted." Kathy Lockett, who lives in England, has found that when she is working at home on her computer and begins wondering what her tabby cat is doing, "within twenty seconds she comes bounding in with her welcome and 'what do you want' call. We chat and then she curls up and keeps me company while I continue working. It has happened over and over in our eighteen years together. I do not have to call her, only think about her, and she appears and talks to me."

Pauline Bamsay, of Port Talbot, Wales, is convinced that her cat has telepathic powers:

> When he is not around, I only have to think "Come on home, Leo," if I feel he's been gone a long time, and within minutes, sometimes under a minute, he will appear, depending on how far away he is. He visits neighbors' gardens and also an old disused allotment area [community garden] just to the rear of our garden. It is his hunting ground. If I am in the garden and thinking "Where are you, Leo?" he calls to me vocally as he approaches the garden. If I am in the house, he comes bounding in through the cat flap in the back door with a loud meow and even comes upstairs to find me. As I write this letter, I can see him on our garage roof, lying curled up asleep. I thought to myself "There you are, Leo." Almost immediately he woke up, stood up, and looked directly at me through the window, which is about 15 feet away from

the garage. Then after a moment he turned and headed toward the allotment across the garage roof!

Some people find that telepathic calls seem to work the other way around: their cats seem to call them. More about this later.

## Dogs

In the training of dogs, whether as working dogs, agility and obedience competitors, or pets, the owners are generally near their dogs, so it is difficult to separate the effects of direct mental influence from the normal senses and training. Most trainers and handlers simply concentrate on performance, without reflecting much on the means of communication between person and dog.

Some working dogs know what to do even when they are at a distance. Raymond McPherson, of Brampton, Cumbria, and his dogs are successful competitors in sheepdog trials, and winners of the International Supreme Championship. He is convinced that his Border Collies are extremely intelligent. But is their skill at anticipating his intentions due more to their intelligence and working routine or to telepathy?

> If you have a dog with a brain and a natural instinct to herd sheep they can do an awful lot on their own without any commands at all. If you go to a sheepdog trial you just see a little of what they can do. It is when you go out on the hills in the day-to-day working with sheep that you find out how intelligent [the dogs] really are. They will work out of sight, some will go three miles away, and one dog can work with anything from three or four to five hundred sheep and sometimes even more. You can build a tremendous connection with them, and they anticipate what you want them to do before you ask them to do it. Kindness is the best way to build a good connection with a dog.

As usual, the best way of teasing apart the influence of sensory clues from direct mental influence is in situations where neither routine nor normal sensory communication can provide a plausible explanation. Some dogs, like cats, respond to silent calls and come to their owners when summoned. Most spectacular are calls when dogs are outdoors and far away from their owners.

Eric LeBourdais trained his Golden Retriever to respond to a dog whistle so that she can range quite far. One day when she was about a year old, he suddenly remembered that he needed to do something at home. She was about a quarter of a mile away. He was about to reach into his pocket for the whistle, but the moment he had the thought she lifted her head as if he had blown the whistle, and headed straight for him. He was surprised, but put it down to coincidence. "However, as the years passed this happened several dozen times—sufficiently often that I am absolutely convinced there was nothing coincidental about it at all. There was never any sort of action on my part—nothing normal that could have accounted for [her response]. Sometimes she would be completely out of sight when I would begin to have thoughts of heading home. Each time she came to me directly, exactly as if summoned."

In the case of working dogs and dogs that respond to calls, the command is usually one with which the dog is already familiar. The telepathy is in the timing. But the most striking instances of silent commands are those in which the person wills the animal to carry out a nonroutine task; the influence in these cases is more detailed and specific. Here is an example of a pet dog responding to the silent commands of his owner, Jane Penney, who lived in Cornwall:

> One day when my dog was sound asleep I deliberately thought to myself, "Wake up and bring me your big blue ball and we'll go and play in the garden." Maggers woke up, went over to his toy bowl and rooted around for the big blue ball (which he didn't like very much), brought it to me, and went to the back door (no, he didn't need to pee). One day, toward the end of his life, he'd left all his toys all over the place and I was falling over them (I'm a bit unsteady, very arthritic). I didn't say a word (the poor dog was asleep) but thought how nice it would be if he could put his toys away. When I came downstairs he was lying with his toy bowl in the middle of the living room floor, with all the toys in the bowl!

## The experiments of Vladimir Bechterev

This kind of telepathic communication was explored by means of experiments by the late Vladimir Bechterev, an eminent Russian neurophysiologist. Although his investigation was carried out many years ago, as far as I know it is still the only experimental study of this subject reported in the scientific literature.

Bechterev, an admirably curious and open-minded investigator, was intrigued by a dog act he saw in a circus in St. Petersburg, in which a Fox Terrier called Pikki seemed to respond to the mental commands of his trainer, W. Durow. Durow told Bechterev that his method was to visualize the task he wanted the dog to do—for example, fetching a book from a table. Then he would hold the dog's head between his hands and look into his eyes: "I fix into his brain what I just before fixed in my own. I mentally put before him the part of the floor leading to the table, then the legs of the table, then the tablecloth, and finally the book. Then I give him the command, or rather the mental push: 'Go!' He tears himself away like an automaton, approaches the table, and seizes the book with his teeth. The task is done."

Bechterev and several of his colleagues found they too could command Pikki in this way, even in the absence of Durow. They carried out a series of trials to find out if they were giving the dog subtle cues by means of eye, head, or other body movements. Bechterev also carried out tests with his own dog, and found that he too could respond to mental commands. He reached three conclusions:

1. The behavior of animals, especially that of dogs trained to obey, may be directly influenced by thought suggestion.
2. This influence may be effective without any direct contact between the sender and the dog who is receiving, as when they are separated by a wood or metal screen or blindfolds.
3. From this it follows that the dog may be directly influenced without the presence of any signs by which he could be guided.[3]

Bechterev regarded these investigations as preliminary and pointed out how desirable it would be to carry out further experiments with dogs: "It would be important to study not only the conditions governing the transfer of the mental influence from the agent to the percipient, but also the circumstances involved in both the inhibition and the execution of such suggestions. This would necessarily be of theoretical as well as of practical interest." Unfortunately the pioneering work of Bechterev has not been followed up, and his words remain as relevant today as when he wrote them more than seventy-five years ago.[4]

## Guide dogs

Some of those who work most closely with dogs are blind people with guide dogs. I wanted to find out if any blind people had noticed that their dog seemed to pick up their intentions without their giving commands either vocally or through body movements. With the help of the British Guide Dogs for the Blind Association, Jane Turney and I asked more than twenty guide dog owners about their experience. We also received much valuable information through letters in response to an appeal for information in *Forward*, a British magazine for guide dog owners, published in Braille.

Some people with guide dogs had not noticed anything of this kind, but most had. Several commented on the fact that it depended on the closeness of their relationship with the dog and that some dogs were much more responsive than others. Even responsive dogs may not pick up their owners' intentions every time. Sarah Craig of Bridgend told us:

> Paxton, a black Labrador, is my second dog. It took me two years to get used to him. Now there are many times that I have felt that he can pick up signals that I do not give consciously, that I don't speak to him. I will think, "We've got to turn right here" and we will turn right, even though I haven't said anything. He picks up very much on things I think and things I feel. There have been times where I have tried to test him, where I have deliberately tried to think about directions in my mind but I have tried to keep as straight as possible, and he has still done it. He doesn't do it all the time, though. Sometimes he is more in tune than others. There are times when he is distracted or I am not thinking clearly enough and not giving him precise directions.

The most striking instances involve the dog responding to thoughts the owners are not planning to put into action right away. For example, Mike Mitchinson was walking with his guide dog from his home in Bath to a particular shop some twenty minutes away. The route took him past his dentist's surgery. "I can remember thinking at the start of the relevant road, 'I must not forget my dental appointment at 10:00 A.M. on Thursday' (this was Monday). I then walked on confidently and only half taking note of where I was. Imagine my surprise when I found myself swinging to the left and entering a gravel driveway! Yes, it was the dentist." Likewise, John Collen of Southend-on-Sea, was walking past some shops one morning and thinking that he would call the grocer that afternoon to get some apples when the dog led him into the grocery store. "I told the owner I was only *thinking* of coming in because I didn't want to carry the apples about, and that I would come back later this afternoon, but the very fact that I thought about it casually was enough for Pedro to pick up on it."

Could it be that the guide dogs are responding to changes in the way the owners walk or hold the harness? Several of our informants had thought about this possibility, including Pedro's owner. "I am totally blind, so I cannot see the dog, and I wouldn't be sure about direction of travel. Under those conditions I wouldn't be making any indication as to direction or stopping or starting, I am just walking along thinking, and that is why I started to believe he is picking up something other than visual cues or other physical indications." Peter Neely, of Kumnock, in Scotland, has come to a similar conclusion:

> When I am working with Sam, the black Labrador I have had as a guide dog for two years, I would definitely say there is a telepathic link there because he seems to know which way I want to go. He seems to anticipate if I am changing my mind on the route. I believe that if you are a guide dog owner, plus a guide dog lover, there is a connection, a sort of invisible umbilical cord between you and the dog because what you are feeling and what he is feeling is traveling up through the harness. Some people might say your subconscious means you do a different tension on to the handle, which the dog picks up, but I honestly don't feel that I am doing that.

Of course these are only opinions, but the opinions of people with years of experience with guide dogs are of considerable value. Given the physical contact through the harness, however, it is difficult to separate telepathic influences from subtle sensory clues, and I have not been able to think of a straightforward experiment with guide dogs that could conclusively eliminate the possibility of unconscious movements by the owner.

## Horses

Many riders experience a close physical, emotional, and mental connection with their horses, which seem to respond to their thoughts. Andrea Künzli of Starrkirch, Switzerland, for example, wrote this: "I can be riding my horse at a walking pace and I think to myself, 'When I get to that tree I'll put her into a trot,' and as if my horse has read my thoughts, and I have at the moment given no (conscious!) body signal, it will start to trot. My husband and daughter have had exactly the same experience with their horses." Experienced riders often take this responsiveness for granted. Here is how Lisa Chambers of Chico, California, a less experienced rider, found it out for herself:

> Riding Kazan became somewhat nerve-racking, since I never knew when he was going to spook. Until, that is, I tried communicating with him telepathically. I first tried it when I wanted him to cross a white wooden bridge. He wouldn't even place a hoof on it the first few times I tried, so the next time I came out to ride, I held in my mind a sharp and clear picture of him walking calmly across the bridge with me on his back. It worked! We approached the bridge, stepped onto it, and crossed it without a moment's hesitation or misstep. Hooray! I was so impressed by the success of my experiment, that I started using telepathy in my daily horse routines. When I want Kazan to step into a horse trailer, I picture it happening and in he goes.

With horses, as with guide dogs, it is difficult to disentangle mental influences from unconscious body signals, such as slight changes in muscular tension. "It is tempting when one is riding a very well schooled horse or a horse that knows one very well, to think the horse is receiving telepathic messages. However, it may just be slight movements of the rider, which are interpreted by the horse and acted upon" (Marthe Kiley-Worthington).[5]

One of the few people to carry out experiments on mental communication with horses was Harry Blake, a British horse trainer famous for his method of "gentling" rather than "breaking" horses. He worked by establishing an empathy with the horse and was often able to train horses remarkably quickly and effectively. His method had something in common with horse whispering and the procedures of the American horse trainer Monty Roberts.[6]

In one series of experiments carried out with a horse called Cork Beg, Blake first trained him to go to one of two food buckets, placed 10 yards apart, directing him "merely by using telepathy" to the one that contained his breakfast, rather than the other, which was empty. "Within a few days he was going straight to the bucket I directed him to, and I persevered with this for a fortnight." In the tests themselves, the horse was offered two buckets containing equal quantities of food. For the first five mornings Blake directed him alternately to the left and the right, then for four mornings to the container on the left.

> The ninth morning brought the most difficult experiment of all. For four mornings running he had taken his breakfast from the container on the left, and on the ninth I wanted to change him to the container on the right. Much to my relief he went straight to it. Having come out of that successfully, he had to take it again from the right-hand container on the tenth morning, from the left on the eleventh and on the twelfth morning from the right. Each morning he went directly to the correct container.[7]

Since the horse could see Harry Blake, it is impossible to rule out subtle visual cues of the kind picked up by Clever Hans. But Harry Blake did other experiments on telepathy from horse to horse, kept in separate buildings out of sight of each other, which seem to rule out such clues from Blake himself or from the other horse. These experiments are discussed in Chapter 9.

## Two-way communication

If invisible bonds exist between animals and people, it would be surprising if they did not permit communications to take place in both directions rather than in just one.

I have collected more than 2,000 accounts of seemingly telepathic or psychic influence of owners on their pets and 371 cases where the influence seems to flow the other way. People seem much less sensitive to these influences than their animals are, or perhaps humans just pay little attention to them. But 371 is still a large number of cases, and presumably many people who have not written to me have had similar experiences.

Out of these 371 cases, 37 concern deaths or accidents in distant places (discussed in Chapter 6). Most of the other cases involve silent calls for help. The majority concern cats.

## Cats calling people

Cats seem especially talented at getting what they want from their owners by subtle means. Some people are convinced that cats can influence them telepathically. The most common situation in which this occurs is when the cat is outdoors and wants to be let in. Here is an example from Sonya Porter of Woking, Surrey:

> My husband, David, soon found that he could tell when Suzie was outside in the garden, wanting to come in. The first time this happened was one Sunday morning when we were in bed, reading the papers. David suddenly said, "Suzie wants to come in," got out of bed, and opened the bedroom curtains to find Suzie sitting on the gatepost, staring intently at the bedroom window. After that, I got quite used to David's going to the front or back door to let Suzie in, even though I never heard her cry or scratch the door. David just said she "fluenced" him.

Some cat owners know not only when cats want to come in but also which of several cats is silently calling them. Laura Meursing kept six cats on her large estate in Belgium: "The cats were often outside on the grounds, but I felt every time when one of the cats wanted to get in and also which one." And a French cat named Minet called his owner telepathically even when she was asleep. Madame G. Woutisseth of Vanves, France, sent this report: "I suddenly know that he is behind the door because his image, in the posture in which I will find him, imposes itself onto me, even waking me up if need be. There is no call, no meowing or other sign. Everything takes place in silence."

Some dog owners have had very similar experiences, like Lydia Arndt, of Riverside, California: "One of my Great Danes can be outside, and when she wants in, she wills me to come to the back door. I can be at the other end of my home and she thinks so hard that I have to stop everything and go let her in. We do this several times a day."

## Calls by lost cats

Cats that roam freely have a tendency to get lost, often because they are unintentionally shut into sheds or garages by neighbors. Some cat owners have found that they can tell where the lost cat is. Solomon, a Siamese cat in Whittlesey, Cambridgeshire, was very inquisitive and tended to get shut up this way. When he did not come home at night, his owner, Celia Johns, had to go and look for him. "I never knew where he was, but I found that if I stood outside the back door and thought hard, I invariably turned in the right direction to find him."

Some stories about the rescue of lost cats are quite dramatic and seem to show that the cat in some way draws the owner toward it. But often this happens only after a frustrating period of searching using the trial-and-error approach. This example came from Martha Lees of Fleetwood, Lancashire:

> In June, Solitaire, the younger of our cats, disappeared. We searched for her without success. The third day, all of a sudden I felt that I had to go out right away. I hurried up the street and turned into Fir Close. I went up to the second house on the right and rang the bell. A gentleman opened the door, I apologized for disturbing him, and told him about my missing cat. He assured me that he had not seen it. I

> asked him if he would mind my having a look around his back garden, as it would make me feel better. He took me through the back and I called, "Solitaire, Solitaire," and immediately a cat started to meow very loudly. I followed the sound to a large mound of garden rubbish. There was a hole at the top and on looking into it I saw my cat's face looking up at me from about three feet down. She was stuck and her neck was bent at an angle. . . . I carried her home joyfully.

In some cases the owners are driving when they seem to know where to go. In the following example, as in the previous one, this knowing did not happen right away but only after exhausting the obvious possibilities. The cat, Whisky, had escaped from a cattery in a Yorkshire village while her family was on vacation. When they returned two weeks later, they learned that she had been missing for almost the entire time they were away.

Her owner, Catherine Forrester, searched the village for her, calling at every house and at the pub. Several people had seen a stray cat, but no one knew where it was: "By the time I had done this it was dark so that it seemed I must go home and return the next morning. Having gone about a mile I felt compelled—no dithering—to turn around and go back to the village. This I did and turned down a dead-end road that leads only to a reservoir. Half a mile or so down, I stopped the car, got out, and called, 'Whisky.' Immediately there was a meow, and she jumped over the wall from a field."

How do cats draw their owners toward them? This phenomenon, which is related to the ability of animals to find their owners, is discussed in Chapter 13.

## Dogs in distress

Most of the stories I have received about dogs influencing their owners from a distance have occurred when the dogs were in great distress. Dolores Katz of Deming, New Mexico, wrote of such a situation:

> One day while at work it started to thunder and rain. As I worked, I became increasingly edgy. Then very agitated . . . Something was

> wrong. At this point I will add that I never took time off from work. I asked my employer if I could take the afternoon off, I didn't feel quite right. On my way home I knew Eric, my German Shepherd, was in trouble, I knew he was bleeding. When I arrived home, I rushed to the back patio. The window was broken. Frightened, Eric had hit the glass with his paw and had sliced off the front pads on the broken glass. He was bleeding very badly. I feel he needed me and called out the only way he could—telepathically—knowing I would come to him.

Some people pick up a feeling of distress without identifying the dog as its source. One day Jill was at work in her office in Exeter when she had a "strange physical feeling" that she could not account for and knew that something was wrong. She felt an urgent need to return home, a mile and a half away, fearing that her elderly mother might be ill. When she reached the house, her mother greeted her with "How did you know?" Their ten-year-old Boxer had had a stroke and was paralyzed. "I'm quite sure that he was making contact with me somehow," Jill said. "He was in a very distressed state and sadly had to be put down soon afterward."

In some cases the distress that people pick up comes from a pet that is actually dying, and I discussed such cases in Chapter 6.

## Horses, cows, and other animals in distress

Dogs and cats are not the only animals that send out distress signals. A Swiss woman who kept sheep told me that one night she woke up and felt she had to go to the barn. When she arrived there she found that one of her sheep had just given birth. She was convinced that the sheep had called her, because she never woke up in this way or went to the barn in the middle of the night.

The horse trainer Harry Blake had a similar experience with a cow. Normally, he said, he slept like a log, but one particular night he woke up with a powerful feeling that something was amiss, so he went out to his animals. He found a cow that was calving. It was a breech delivery, and she was in difficulty. On thinking about it afterward he worked out that he'd been awakened by the feeling that something was wrong and had been drawn subconsciously to where the cow was. He had similar experiences with horses. On one occasion, a horse to which he was very attached woke him up at three in the morning. "I simply knew there was something wrong, and when I went out to have a look at him I found that he was having a violent attack of colic."[8]

Other people have not only picked up a horse's distress but have gained more specific information as well. Charles Craig, for example, woke up one night feeling uneasy and apprehensive. He got dressed, went downstairs, and picked up his wire cutters and a torch. Then he put on his boots and went out, on a very dark night, directly to the very spot, about half a mile from his house, where his favorite mare was caught in barbed wire in a bog. He said that as he went downstairs he "knew exactly where the mare was and exactly what had happened" because he could "see it in his mind's eye."[9] Dagmar Bruns-Jensen of Muhlenhaus, Germany, wrote: "A few months ago I woke at night, and half asleep and half awake I saw my horse, Kenia, who could not move because the blanket was wound around her legs. First I decided to fall asleep again, but it didn't work. The picture of Kenia came again and again. So I decided to go down to the stable, and Kenia was standing there shackled by the blanket around her legs."

Sometimes people respond to distressed animals that they do not even know. Lucy Crisp, for example, was asked to feed a neighbor's cats while the neighbor was on vacation. The first day she felt uneasy when she was leaving the house. The second day her uneasiness became more intense, and she went around to the back of the neighbor's house where she found a cage containing two desperate rabbits that had not been fed or given water for several days. "I have no idea how these animals managed to send out their distress signals," she said.

## Animal communicators

In addition to these communications between domestic animals and their owners, there is a long tradition of communication with animals by shamans in tribal societies. One of the powers attained by yogis in India was said to be an ability to understand the signals of all kinds of animals. Some of the most appealing Christian saints, like Francis of Assisi and

Cuthbert, were said to communicate with animals and to understand their language. And there have always been animal keepers and trainers who are extraordinarily in tune with their animals and seem to know what they are feeling. In fiction, stories about Dr. Doolittle and others like him have a deep appeal.

There are also people who make a living as "animal communicators," who claim to pick up telepathically what people's pets are thinking and feeling. Some give counseling and advice for a fee, either in person or over the telephone.

Given the background of shamanic communication with animals, the telepathic experiences of pet owners with their pets, and the remarkable sensitivity that some people have to animals, I am happy to accept that some animal communicators may indeed have extra-ordinary powers, even if these do not usually lend themselves to scientific testing. However, many of these so-called animal communications, especially when carried out for profit, may well be a projection of the communicator's own thoughts rather than genuine cases of telepathy.

Animal communicators themselves are well aware of the problems. Penelope Smith of Point Reyes, California, who has trained hundreds of people in "interspecies telepathic communication" in her workshops, has seen people "mixing their communication abilities with their own agendas or emotional shortcomings." She has proposed a code of ethics for interspecies telepathic communicators, which includes the following passage: "We realize that telepathic communication can be clouded or overlaid by our own unfulfilled emotions, critical judgments, or lack of love for self and others."[10]

Professional animal communicators are often ready to offer information about animals' feelings and even about their past lives, and they may well play a valuable role in counseling the animals' owners. But they are often reluctant to provide information that can be more immediately verified. In his book on this subject, *Communicating with Animals,* the veteran reporter Arthur Myers[11] explains how he interviewed many communicators to find out how successful they had been at finding lost animals telepathically. Most told him they try to avoid such jobs. However, Myers did find some cases where communicators had been able to locate lost animals by describing where they were and by giving clues that enabled them to be found.

For me, the most interesting of these apparent communications from animals are those that can be tested empirically. I agree with Myers that the finding of lost animals seems the best place to begin.

## Telepathic telephone calls

Before the invention of modern telecommunications, telepathy would have been the only way people could reach each other at a distance. In Chapter 4, I mentioned a story by Laurens van der Post about the way the Bushmen of the Kalahari Desert knew what members of the group were doing many miles away and when they were returning. The Bushmen were under the impression that the white man's telegraph also involved a kind of telepathy.

Even in traditional long-distance communication by drumming, the message may not be transmitted simply through the sounds. Richard St. Barbe Baker suggested in his book *African Drums* that the drumming may serve primarily to tune in the senders' and receivers' attention to each other: "May it not be that the drums create the atmosphere for the transmission of thought messages and vision which annihilate time and space? The more deeply I have delved into the problem of transmission, the more I have become convinced of the inseparable association between the transmission of a visual picture by telepathic means and the language of the drum."[12]

People in modern societies are not encouraged to develop telepathic skills. They are regarded as superstition by rationalists and ignored by institutional science and the educational system. In any case, modern technology usually provides much easier and more effective means for communication at a distance. Television allows everyone to see images from far away, and telephones provide instant worldwide communication.

But telephones also provide an excellent opportunity to study telepathy, precisely because they fulfill the same function: enabling people to communicate at a distance. In order to make a telephone call to someone it is necessary to *intend* to call that person. The very act of calling someone focuses attention on that person at a distance. We have already seen that pets can respond to their owners' calls and intentions at a distance. Can some pets tell when their owners are calling, even before the telephone is answered?

## Telephone-answering cats

I have received fifty-nine reports of cats that respond to the telephone when a particular person is calling, before the receiver is picked up. In all cases, the person calling is someone to whom the cat is deeply attached. In Veronica Rowe's case it was her daughter Marian, whose cat, Carlo, would not allow any other member of the family to cuddle her.

> Seven years after she acquired Carlo, my daughter went to teacher training college and rang us infrequently. However, when the phone did ring and it was Marian . . . Carlo would bound up the stairs (the phone was on the half landing) before I had picked up the receiver. There was no way that this cat could have known my daughter was to ring us. It was a standing joke when he bounded up the stairs that Marian was on the other end of the line. He never did this at any other time and was not allowed upstairs anyway.

Godzilla lives with David White (Figure 8.1), a public relations consultant who works from his home in Watlington, near Oxford. He used to go away several times a year, and his parents would come to watch the house, look after the cat, and answer the many telephone calls. David would call home from North Africa, the Middle East, and continental Europe to check that all was well and to pick up any messages. "Whenever I called, my cat would run and sit beside the telephone, as it was ringing," David reported, "whereas she ignored the other calls my parents took on my behalf. And the calls were made at random times." Godzilla responded in this way before the telephone had been answered, so he could not have been reacting to David's voice.

Most of the cats that were said to respond to telephone calls from particular people reacted when the telephone began to ring, but five did so even before the ringing began. For example, Helena Zaugg describes how her family's cat in Brügg, Switzerland, responded to her father, to whom it was closely attached:

***Figure 8.1** David White and his telephone-answering cat, Godzilla, in Watlington, Oxfordshire. (Photograph: Phil Starling)*

After my father had retired he sometimes worked for an acquaintance in Aargau. Sometimes he called us from there in the evening. One minute before this happened the cat became restless and sat down next to the telephone. Sometimes my father took the train to Biel and then used a moped to get home from there. Then the cat sat down outside the front door thirty minutes before he arrived. At other times he arrived at Biel earlier than usual and then called us from the station, and the cat sat down near the telephone shortly before the call came. Afterward she went to the front door. All this happened very irregularly, but the cat seemed to know exactly where he was and what would happen afterward.

Likewise, a Siamese cat belonging to Vicki Rodenberg "perked up when a particular person called on the phone—only she did it moments before the phone rang! She would actually run to the phone and cry, and [it was] always that person."

I have also received reports of cats that come to the telephone immediately *after* it has been answered when a particular person is ringing, but in such cases it is impossible to know whether the cat was simply responding to the sound of the person's voice or to the reactions of the person answering the telephone. I have therefore excluded these cases from consideration.

## Dogs and telephones

Many dogs bark or react in some other way to the telephone when it is ringing, whoever is on the other end. But we have sixty-six cases on our database where dogs respond to the telephone only when a particular person is calling and before the telephone is answered. As in the case of cats, the people to whom the dogs reacted were their owners or other members of the family to whom they were particularly attached. Poppet, for example, was very attached to Margaret Howard's mother and knew when she was telephoning or coming to visit:

> I first noted that Poppet showed restlessness, excitement—ears pricked up, tail wagging, wandering between front and back doors—and she developed a special type of bark, which I called yipping, and surely within minutes my mother arrived. No special times or routine to her visits, but Poppet's reaction was always the same, morning, noon, or night. I gradually got to notice that I could tell whether my mother was coming via the front or back door, as Poppet would position herself at the right one. I also noted that when the telephone rang, although Poppet would look at it, she did not particularly bother, but I always knew when my mother was on the phone, as Poppet appeared all excited, standing by the phone and using her special yipping call.

As in the case of cats, this response of dogs does not seem to diminish with distance. Marie McCurrach of Ipswich, Australia, had a Labrador that joined the family when her son was ten. Four years later, her son went away to a naval school in north Wales and then went on to serve in the merchant navy, mainly sailing to and from South Africa. His mother recalls that "Every time he rang home the dog would run to the telephone before anyone could answer it. The dog never bothered about any other calls, only our son's, and we had to hold the phone to the dog's ear so that our son could speak to him and the dog would reply. Our son never gave us a time that he would phone and did not phone on the same day of the week or anything like that, so how did the dog know it would be our son on the phone before anyone had lifted the receiver?"

Some dogs were said to react before the telephone actually rang. When Tam, who lived in England, was traveling abroad for six months, she left her German Shepherd with her parents: "My mother used to say that before I called her the dog would actually bark at the phone and run to and fro, between her and the phone. He would then sit in front of the phone and watch it. A few moments later I would call." A dog called Jack belonged to a family living near Gloucester, where the father worked for the Ministry of Defense. Some nights he could not return home owing to the urgency of the work or the lateness of the hour, and on these occasions he called to say so. "Ten minutes or so prior to that call being received, the dog would sit by the telephone until it rang," writes his son, Mr. S. Waller. "On the nights when no telephone call had to be made by my father the dog made no move whatsoever, and remained in his basket. Further, the dog would pay no attention to any other telephone calls at any other time."

Dogs and cats are not the only pets that seem to know in advance who is calling. Some parrots do so too, and so did a pet capuchin monkey called Sunday. Her owner, Richard Savage, left the monkey with a friend in British Columbia, Canada, while he was away on filmmaking assignments. According to this friend, "Several minutes before Richard telephoned to talk to me, Sunday would jump up and start chattering. After his call, she would settle down for days, ignoring the telephone—until just before Richard called again."[13]

## People who know when a certain person is calling

Once when I was discussing this research at a seminar, someone asked, "If cats and dogs know when particular people are calling, then why don't people have this ability?" A good question. I then remembered that I myself had thought of particular people for no obvious reason, and then, shortly afterward, they called. I asked the participants in the seminar if anyone else had noticed this phenomenon, and to my surprise almost everybody had.

My colleagues and I have now found out through a series of surveys in Europe and North America that most people have experienced telephone telepathy. For no apparent reason when they think about a particular person, then the phone rings and that person is on the line. Or else when the telephone rings they have an intuition about who is calling, and they turn out to be correct. Here are a few examples taken from hundreds on my database:

- "On many occasions my mum phones me at the same time as I phone her."
- "In the last few months almost every time I think of Belinda she rings me."
- "My elder daughter and I more often than not will say, 'I was just thinking about you,' when we speak to each other on the phone."

Such experiences are the most common kinds of apparent telepathy in the modern world.[14] They generally occur between people who are closely bonded, like family members and best friends.[15]

Most of the unexplained powers discussed in this book are better developed in animals than in people. Normally dogs seem to be the most sensitive, followed by cats, parrots, and horses, with humans trailing far behind. But here, for a change, is an ability that seems better developed in people than in animals.

But even if most people have had seemingly uncanny intuitions about who is telephoning, is a mysterious psychic power really at work, or can it all be explained as an illusion? Could apparent telephone telepathy be a matter of coincidence? Perhaps people often have thoughts about others for no particular reason. By chance these thoughts may be followed by a telephone call from that person. If people only remember the times they are right and forget the times they are wrong, then an illusion of telepathy may be created by a combination of coincidence and selective memory.

Alternatively, a person may be expecting a call at a particular time from a particular person but may be unconscious of this expectation. So when the call comes there is no need to invoke telepathy because an unconscious expectation could explain it instead. The trouble is that an unconscious expectation is elusive. Indeed this may be an untestable hypothesis, because if the expectation of the telephone call is unconscious, how can anyone prove that it is really there? And if it is really there, then might it be a *result* of telepathy rather than an alternative to it?

The best way to resolve these questions is by means of experimental tests.

## Experiments on telephone telepathy

I have developed a simple experimental procedure in which participants receive a call from one of four different callers. They know who the potential callers are, but they do not know which one will be calling in any given test, because the caller is picked at random by the experimenter. They have to guess who the caller is before the caller says anything. By chance they would be right about one time in four, or 25 percent of the time. Are they right significantly more often than would be expected on the basis of random guessing?

My colleagues and I conducted more than a thousand trials of this kind. The average success rate was 42 percent, very significantly above the chance level of 25 percent, with statistical odds against chance of billions to one.[16] We also carried out trials in which the participants were filmed continuously to rule out the possibility that anyone was cheating by getting messages by e-mail or by unscheduled telephone calls, and in the filmed trial the hit rate was even higher, 45 percent, and very significant statistically. I also carried out an experiment for a British television program, starring a girl band called the Nolan Sisters, whose hit rate was 50 percent, or twice the chance level.[17]

We carried out another series of trials in which two of the four callers were familiar, and the other two were strangers whose names the participants knew but whom they had not met. With familiar callers the success rate was more than 50 percent, highly significant statistically. With strangers it was near the chance level, in agreement with the observation that telepathy typically takes place between people who share emotional or social bonds.[18]

In addition we found that these effects do not fall off with distance. We recruited participants in Britain who had recently arrived from Australia or New Zealand, and we did tests in which some of their callers were in Britain and others back home, literally the other side of the world. The subjects identified their nearest and dearest thousands of miles away more successfully than new acquaintances in Britain, showing that emotional closeness is more important than physical proximity. The details of these experiments have been published in scientific journals and can be read online on my website (www.sheldrake.org). Our positive findings in telephone telepathy tests have now been replicated at the universities of Amsterdam, Holland,[19] and Freiburg, Germany.[20]

Telepathy continues to evolve. One of its latest manifestations is the telepathic e-mail. People think of someone who shortly afterward sends them an e-mail. We have done more than 700 tests on e-mail telepathy, following a similar design to the telephone tests, with an average hit rate of 43 percent, again very significantly above the chance level of 25 percent.[21] A similar phenomenon occurs with SMS messages.[22] Again the results have been published in peer-reviewed journals and are available on my website.

The latest experiments of this kind are automated telephone telepathy tests that work through mobile phones, making it possible for anyone to take part wherever they are. The results so far are way above chance and highly significant statistically. The details are given on my website at the Online Experiments Portal.

Although telephones have superseded telepathy for most practical purposes, they are also helping us to rediscover it.

Chapter 9

# Animal-to-Animal Telepathy

If telepathy occurs between animals and people, and from people to people, what about telepathy from animals to animals?

Wild animals that live in social groups are often strongly bonded to one another and practically incapable of leading an isolated existence. Complex social organization occurs even among the lowest animals, such as corals and sponges. In fact, what we recognize as a coral or sponge is a colony of millions of tiny organisms that together form a kind of superorganism, with its own characteristic form.

In this chapter I discuss the way that societies, schools, flocks, herds, and other social groups are organized. The activities of the individual animals within a group are coordinated through the group's field. We have seen how this social field, called a morphic field, links humans and animals together and provides a means of telepathy between pets and their people. The same kind of bonding occurs between animals in the wild, and it is in this bonding that the roots of animal-to-animal telepathy lie.

I begin by considering some of the most complex forms of social organization in the animal kingdom, achieved by creatures with brains smaller than the head of a pin.

## Social insects as superorganisms

Societies of termites, ants, wasps, and bees may contain millions of individual insects. They build large and elaborate nests, exhibit a complex division of labor, and reproduce themselves. They have often been compared to organisms or superorganisms. For example, two of the world's leading experts on ants, Edward O. Wilson and Bert Hölldobler, wrote in their book *The Superorganism* as follows:

> Consider one of the most organism-like of insect societies, the great colonies of African driver ants. Viewed from afar the huge raiding column of a driver ant colony seems a single living entity. It spreads like the pseudopodium of a giant amoeba across 70 meters or so of ground. . . . The swarm is leaderless. . . . The frontal swarm, advancing at 20 meters an hour, engulfs all the ground and low vegetation in its path, gathering and killing almost all the insects and even snakes and other larger animals unable to escape. After a few hours the direction of flow is reversed, and the column drains backward into the nest holes.[1]

Social insects and all social animals are bound together in their social groups by morphic fields, which carry the habitual patterns and "programs" of social organization. These fields coordinate the architectural activities of social insects that build nests and other structures. The fields contain, as it were, an invisible blueprint for the nest. The morphic field of the colony is not merely inside the individual insects; rather they are within the morphic field of the group. The field is an extended pattern that contains all the individual insects, just as the gravitational field of the solar system contains the sun and the planets and coordinates their movements.

Much has been discovered about the ways in which social insects communicate, through shared food, scent trails, touch, and vision—as in the wiggle dance of honeybees, through which they communicate the direction and distance of food. But all these forms of sensory communication work together with their connections through the morphic field of the group. It is this field, I suggest, that enables the insects to interpret the scent trails, dance patterns, and so on, and to react appropriately.

Sensory communication by itself would be totally inadequate to explain how termites, for example, could build such prodigious structures, with nests up to 10 feet high, filled with galleries and chambers and even equipped with ventilation shafts. These insect cities have an overall plan that far exceeds the experience of any individual insect.

Karl von Frisch, who discovered the wiggle dance of bees, has written an excellent book on animal architecture in which he discusses the complex buildings of termites. The insects are blind and cannot see one another, but they mark trails with scent so that other termites can follow them, and they give knocking signals by striking a hard surface with their heads. But von Frisch points out that "The information content of both modes of communication is small. The scent trail may lead to a goal, but it cannot explain what should be done there. Drumming is an alarm signal by which soldiers or workers induce other workers to flee into the interior of the nest. . . . But it is just a general warning signal." He concludes that the "finished structures seem evidence of a master plan that controls the activities of the builders and is based on the requirements of the community. How this can come to pass within the enormous complex of millions of blind workers is something we do not know."[2]

Fortunately, a crucial experiment that sheds some light on this question has already been carried out by the pioneering South African naturalist Eugene Marais. He started by observing the way in which workers of a *Eutermes* species repaired large breaches he made in their mounds. The workers started repairing the damage from every side, each carrying a grain of earth coated with its sticky saliva, and gluing it in place. The workers on different sides of the breach did not come into contact with one another and could not see one another, being blind. Nevertheless the structures they had built out from the different sides did join together correctly. The repair activities seemed to be coordinated by some overall organizing structure, which Marais attributed to the "group soul." I think of it as a morphic field.

He then carried out an experiment to see what happened when the termites repairing the breach were separated from one another by a barrier. He divided the termitarium into two separate parts with a steel plate. Now the builders on one side of the breach could know nothing of those on the other side by sensory means: "In spite of this the termites build a similar arch or tower on each side of the plate. When eventually you withdraw the plate, the two halves match perfectly after the dividing cut has been repaired. We cannot escape the ultimate conclusion that somewhere there exists a preconceived plan, which the termites merely execute."[3]

Unfortunately no one has ever repeated this experiment or other experiments of Marais's, which also seemed to show that the members of the colony were linked together by an "invisible soul." I believe that this area is full of potential for fertile research.[4] If the behavior of social insects is coordinated by a kind of field so far unrecognized by biology and physics, experiments with social insects could tell us something about the properties and nature of such fields, which may well be at work at all levels of social organization, including our own.

Deborah Gordon and her colleagues from Stanford University have studied harvester ants in the Arizona desert, and their unexpected and intriguing observations suggest another possible line of research. In order to study how ants switch tasks within the colony, the researchers collected ants so they could mark them with paint, before releasing them and studying which tasks they performed on subsequent days. To collect them they used an aspirator, a tiny vacuum device that quickly sucked the ants into a tube and deposited them in a vial. "Collecting foragers was easy; foragers didn't seem to react at all when their fellow foragers suddenly disappeared from the trail. We were just another predator, like the horned lizard that stands beside the trail sucking up ants while the surviving foragers keep walking by, apparently heedless."[5] Collecting midden workers was easy too, and so was collecting nest maintenance workers, as long as the aspirator was operated properly. If any air flowed out of it, the gust of smells sometimes sent the maintenance workers running back to take cover in the entrance to the nest.

But collecting patrollers was completely different: Even the most careful aspirating of only a few patrollers, well apart from one another, could cause the whole colony to shut down for the day. At first Gordon thought that this must be because of the change in pattern of interaction between the patrollers with one another and with other ants. But it happened too quickly: "When some patrollers outside the nest disappeared, the rest of the patrollers sometimes headed back into the nest immediately, within seconds—long before there was time for anyone to go back into the nest and assess the rate at which patrollers were returning. This happened when we put the aspirator right over the patrollers . . . so that there would be little opportunity for a cloud of alarm pheromone to escape. The patrollers still reacted, often at a distance that seemed too great for any pheromone to travel so quickly."[6]

Gordon then explored the possibility that this reaction depended on the rate at which patrollers met one another and touched antennae. But simulations showed that the variation in intervals between interactions was so large that "it would be hard for an ant to reliably detect a fall in the rate of interactions. So I don't know how to account for the reaction of the patrollers."[7]

Probably the most decisive experiments to distinguish between the field approach to animal societies and the conventional approach could be done with ants and termites.[8]

## Schools of fish

At a distance, a school of fish resembles a large organism. Its members swim in tight formations, wheeling and reversing in near unison. "Either dominance systems do not exist or are so weak as to have little or no influence on the dynamics of the school as a whole. When the school turns to the right or the left, individuals formerly on the flank assume the lead."[9] When under attack, a school may respond by leaving a gaping hole around a predator. More often the school splits in half and the two halves turn outward, swim around the predator, reverse direction, and eventually rejoin each other. This "fountain effect" leaves the predator ahead of the school. Each time the predator turns, the same thing happens.

The most spectacular of the school's defenses is the so-called flash expansion, which on film looks like a bomb bursting. All of the fish simultaneously dart away from the center of the school as a group is attacked. The entire expansion may occur in as little as one-fiftieth of a second. The fish may accelerate to a speed of ten to twenty body lengths per second within that time. Yet the fish do not collide. "Not only does each fish know in advance where it will swim if attacked, but it must also know where each of its neighbors will swim."[10] This behavior has no simple explanation in terms of sensory information from neighboring fish because it happens far too fast for nerve impulses to move from the eye to the brain and then from the brain to the muscles.

Even in normal schooling behavior, it is not clear how the movements are coordinated. Fish can continue swimming in schools at night, so their movements do not seem to depend on vision. There have even been laboratory experiments in which fish were temporarily blinded by being fitted with opaque contact lenses. But they were still capable of joining and maintaining their position indefinitely within the school. Perhaps they could judge the position of their neighbors by their pressure sensitive organs, known as the lateral lines, which run along their length. But in other laboratory experiments by fish researchers, this idea has been tested by cutting the nerves from the lateral lines at the level of the gills. Such fish still school normally.[11]

Even if we understood the means by which they are aware of one another's position through their normal senses, we would not be able to explain their rapid responses. A fish cannot possibly sense in advance where its neighbors are going to move.

But if the behavior of schools is coordinated by morphic fields, then these links and connections become easier to understand, in principle. The field helps shape the behavior and activity of the school as a whole, and the individuals within it respond to their local field environment.[12] A simple physical analogy is provided by iron filings in a magnetic field. When the magnet is moved, the iron filings take up new positions and make new patterns of lines of force. This happens because each individual filing is responding to the field within and around it, and the field as a whole shapes the overall pattern.

It would be fascinating to know what would happen if two parts of a school of fish were separated from each other by a barrier that blocked normal sensory contact. Would their activities continue to be coordinated? As far as I know, no one has yet attempted this kind of research.

## Flocks of birds

Flocks of birds, like schools of fish, show such remarkable coordination that they too have often been compared to an organism. The naturalist Edward Selous wrote of the movement of a vast flock of starlings: "Each mass of them turned, wheeled, reversed the order of their flight, changed in one shimmer from brown to gray, from dark to light, as though all the individuals composing them had been component parts of an individual organism."[13] Selous studied the behavior of flocks of birds over a period of thirty years and became convinced that it could not be explained in terms of normal sensory communication: "I ask how, without some process of thought transference so rapid as to amount practically to simultaneous collective thinking, are these things to be explained?"[14]

There has been surprisingly little research into the behavior of flocks, but in a landmark study in the 1980s by Wayne Potts, the banking movements of large flocks of dunlins were studied by taking films with very rapid exposures, so they could be slowed down to study the way in which the movements of the flocks occurred. Analysis revealed that the movement was not exactly simultaneous but started either from a single individual or from a few birds together. This initiation could occur anywhere within the flock, and the maneuvers always passed through the flock like a wave radiating from the site of initiation. These waves moved very rapidly and took on average only 15 milliseconds (thousandths of a second) to pass from neighbor to neighbor.

In the laboratory, captive dunlins were tested to find out how rapidly they could react to a sudden stimulus. The average time they took to show a startle reaction after a sudden flash of light was 38 milliseconds. This means that it is impossible that they could bank in response to what their neighbors do, since this response occurs much quicker than their minimum reaction time.

Potts concluded that birds respond to a "maneuver wave" that passes through the flock, adjusting their flight pattern to anticipate the arrival of the wave. He has proposed what he calls the chorus line hypothesis to explain this phenomenon, based on experiments carried out in the 1950s with human chorus lines. The dancers rehearsed particular maneuvers, and in some experiments these maneuvers were initiated by a particular person without warning, and the rate at which they passed along the line was estimated from films. This was on average 107 milliseconds from person to person, nearly twice as fast as an average human visual reaction of 194 milliseconds. Potts suggests that this was due to the individual seeing the approaching maneuver wave and estimating its arrival time in advance.

In other words, Potts sees the birds or the dancers reacting to the maneuver wave as a whole. They reacted not to other individuals so much as the spreading pattern itself. This looks very like a field phenomenon, and I suggest that the maneuver wave is a pattern in the morphic field. This seems to me a more plausible explanation than the alternative, that the whole wave is coordinated through purely visual stimuli. This would require birds to be able to sense, notice, and react to such waves almost immediately, even if they are coming from directly behind them. This would require them to have practically continuous, unblinking 360-degree visual attention. The field hypothesis would make it easier to understand how the birds not only perceive and respond to the maneuver wave as a gestalt, a combination of form and wholeness. Through it they could grasp the movement of the flock as a whole and respond to it in accordance to their position within it. The field underlies the continuum of the flock and the movement of patterns through it.[15]

Since the 1980s there have been many attempts to model the behavior of flocks of birds and schools of fish on computers. Craig Reynolds developed one of the best known of these models, called boids, in the 1980s.[16] This model was two-dimensional, but at first sight it seemed to simulate flock behavior quite impressively. The boids model was "individual based," that is to say that it started from individual boids. These boids were programmed to behave according to three simple rules:

1. Steer to avoid being too close to neighbors.
2. Steer toward the average direction that neighbors are heading in.
3. Steer to move toward the average position of neighbors.

By following these rules, a collection of boids on the computer screen behaves rather like a flock. This simulated behavior seems to show that the behavior of the flock as a whole is a product of individuals interacting with their neighbors according to simple rules, with no need for any mysterious organizing principles. But while this may be true of the computer model, it bears little relation to the behavior of real, three-dimensional flocks of birds. Reynolds's boids program started not from data about real birds but rather from a kind of computer programming involving two-dimensional models in which neighboring units interact according to simple rules. Special-effects wizards have used these kinds of programs to create animations of flocks or herds in films like *The Lion King* and *Batman Returns*.

Computer models of the boids type are useful for producing two-dimensional animations, but they are biologically naive. Although there has been little research into the behavior of real flocks, enough is already known to rule out the kind of neighbor-to-neighbor interactions on which the boids-type models depend. Attempts to make mathematical models are now more sophisticated. One class of models combines the principles of magnetism and hydrodynamics. In these models each bird is like a magnetic domain within an iron magnet. Neighbors influence one another through their fields, and the flock as a whole has a field, like a magnet. But unlike a magnet the flock is in motion, polarized in a particular direction, and by introducing the principles of fluid flow it is possible to make models of some of the features of bird flocks, fish schools, and animal herds in motion.[17]

Other models consider what happens if a few members of the flock or school of fish have a preferred direction, for example, based on knowledge of some feature in the environment not shared by other members of the flock. In these models a minority of individuals has a strong preference, while most members of the flock do not. By moving in their preferred direction these few motivated individuals can cause the whole flock to move with them, if they all tend to move in the same direction. But if some of the motivated minority wants to move in one direction and others want to move in another, the models show that the outcome depends on a kind of majority vote. If the numbers of birds motivated to go in different directions are approximately equal, the flock may split into two. But if one direction has a majority in its favor, this majority will predominate and the flock will move in that direction, despite the minority that wants to go the other way.[18]

These models make it clear that groups are unlikely to remain cohesive and allow the spread of information if individuals respond only to others very close to themselves. "As sensory range is increased, a response to a greater number of neighbors increases cohesion and allows effective long-range transfer of directional information."[19] But although models based on strong preferences by a minority of birds may explain some aspects of flock movement, they cannot explain them all, especially when the birds are not moving in search of food or any other directional goal but simply flying together as starlings do before roosting.

A recent study looked at the behavior of huge flocks of starlings containing up to 2,600 birds flying above the Rome railway station. Starlings habitually gather in the evening and fly together over the roost for up to twenty minutes in spectacular aerial displays. They form sharp-bordered, strongly cohesive flocks. A team of scientists, including engineers and physicists, set out to measure the way in which the birds moved by setting up two cameras at right angles to each other, and they analyzed the digital films using a computer program that could track individual birds as they moved within the flocks. This was a complex technical challenge, and none of their films lasted more than eight seconds because of the limited storage capacity of their recording devices. Also, because the cameras were fixed, they could film starlings only in a limited range of movement, when they were flying almost horizontally. They could not observe other aspects of flock behavior, such as the response of the flock to an attack by predators.

A detailed mathematical analysis of the data showed that individual birds were influenced by their neighbors up to seven birds away. What mattered was the number of neighbors rather than their distance from any particular bird. This means that flocks of high and low density showed similar patterns of influence by neighbors.[20] Another important finding was that when the flock turned, the individual birds all changed direction within the flock, rather than wheeling in parallel paths. For example, in a 90-degree left turn, birds at the front of the flock ended up at the right of it, while those on the left ended up at the front.[21]

The more that flocks of starlings are studied, the more remarkable their behavior seems. In a series of observations on flocks of starlings near the center of Rome, published in the *Proceedings of the National Academy of Sciences* in 2010, it seemed that every bird in the flock was influenced by every other bird, however large the flock. In technical language, the birds were said to show "scale-free correlations," meaning that "the behavioral change of one individual influences and is influenced by all the other individuals in the group. . . . The effective perception range of each individual is as large as the entire group . . . making the group respond as one." The authors of this study concluded: "How starlings achieve such a strong correlation remains a mystery to us." They suggested that their observations supported the metaphor of a collective mind.[22]

I suggest that this collective mind is based on the morphic field of the flock. But this collective field linking the birds may operate not only when the birds are flying together but also when they are engaged in other activities. For example, when birds are foraging, if some members of the group find a good source of food, news of this discovery could pass through the field of the scattered flock to other members and perhaps also set up an attraction so that the birds can go in the right direction.

At least one naturalist, William Long, has observed that birds do indeed seem to respond in this way to the finding of food. He fed wild birds at irregular intervals and noticed that when some found the food, others nearby soon appeared. There is no mystery in this, since they could have seen or heard the birds that were feeding. But he also found that relatively rare birds, widely dispersed over the countryside, would rapidly appear when food was available. After many observations, he came to the conclusion that the reasonable explanation was either that feeding birds can send forth a "silent food call" or that their excitement spreads outward. He suggested that it is "felt by other starving birds, alert and sensitive, at a distance beyond all possible range of sight and hearing."

To follow up such observations experimentally, it should be possible to work with flocks of domesticated birds such as chickens, ducks, and geese. Two parts of a flock could be separated from each other so that no influence could pass by through the normal senses. If one part of the flock is frightened or disturbed, is an influence communicated to the other part? If one part of the flock is fed, does the other part become excited at the same time?

## Telepathy within herds

Naturalists and hunters who have studied the behavior of herds of wild animals, including caribou and elk, have often observed that a whole herd can get alarmed and flee after one member senses danger. In some cases this can be explained in terms of sensory signals, but in others observers are often at a loss to explain the sudden flight of animals that shortly before, under the same circumstances, were feeding or resting without suspicion. A sense of danger or alarm can spread silently and rapidly.

William Long studied the reactions of caribou on many occasions and in considerable detail. On one occasion after he had been tracking a herd for hours in New Brunswick, from the trail he read that one member of the herd was wounded, walking on three legs, his right forefoot swinging helpless as he hobbled along. Finally Long came to a wooded slope from which he actually saw the herd about a mile away through his binoculars. As he began to approach them, keeping well out of sight, he came upon the solitary trail of the cripple and shortly afterward startled this animal in a thicket. It hobbled away into the woods. Long found an opening in the cover and turned his binoculars on the other caribou. Already they were in wild alarm and running off rapidly. He was convinced the rest of the herd could not hear, see, or smell him. He was still far away and yet they reacted immediately after the cripple was startled "as if he had rung a bell for them." Long followed the trail of the herd back to where they had been resting before taking alarm, and found there was no track of man or beast in the surrounding woods to account for their flight. He concluded that they had received some silent warning.

This does not always happen. Sometimes a member of the herd can be surprised without alarming the others. Long thought that in this case the solitary caribou was tremendously startled and may have given a particularly intense warning to the rest of the herd. Similar observations of the behavior of elk led him to conclude that whole herds could suddenly feel and understand the silent impulse to flee, and obey it without question in a way that was essentially telepathic.[23]

## Experiments with horses

The British horse trainer Harry Blake was convinced that horses were in communication with one another telepathically and could also respond telepathically to people. He believed that this kind of communication was vital for their survival, since in the wild a herd of horses may well be scattered with some members out of sight and sound of one another: "If one part of the herd should be frightened by the appearance of man, wolf, or some other predator, the rest of the herd, maybe among the trees, can be alerted by ESP even though they can neither see nor hear their fellows. Horses thus alerted will become first disturbed, then prick up their ears and snort, and start to move away from the area."[24]

Blake carried out a number of studies of telepathy between horses. He chose pairs of horses that were brothers or sisters and living as close companions, in the habit of grazing together, walking together, and acting together.

He separated the pair and kept them out of sight and hearing of each other. Then one of each pair was fed and the other observed. For the purpose of this experiment the horses were not fed at the same time each day, nor were they fed at their regular times. In twenty-one out of twenty-four such tests, Blake observed that when one horse was being fed, the second horse became excited and demanded food, though it could not see or hear the first one.

In a further series of experiments, when one of the pair of horses was taken out and exercised, on most occasions the other horse became excited. In another type of experiment Blake made a fuss over one of the pair, usually the one he liked least, and in most cases the other showed signs of disturbance, suggesting it was jealous.

Overall, Blake carried out 119 experiments, and the results were positive in 68 percent of them. He also ran a control experiment with a pair of horses who were hostile to each other. In only one out of fifteen experiments was there a positive result.

As far as I know, these pioneering experiments have never been repeated. But they show that telepathic communication between horses and between other animals can be investigated by means of simple, straightforward experiments.

## Experiments with dogs and rabbits

The only experiments I know of to test for dog-to-dog telepathy were carried out with Boxers by Aristed Essner, a psychiatrist at Rockland

State Hospital in New York. His research was prompted by rumors that Soviet scientists were testing animals for ESP. One of these stories was that baby rabbits had been taken on board a submarine, while their mother was kept in a laboratory on land. When the vessel was submerged, the babies were killed one by one. The mother was said to have become agitated at the very moments they were killed.[25]

For his experiments, Essner used two soundproof rooms in different parts of the hospital. A mother Boxer was kept in one of these rooms and her son in the other. These dogs had been trained to cower at a raised rolled-up newspaper. In the experiment, the son was "threatened" by an experimenter with a rolled-up newspaper, and he cowered. The mother, in her isolated chamber, cowered at exactly the same moment.[26] In another experiment, a Boxer was kept in one of the chambers and wired up to an electrocardiograph while his owner was in the other chamber. The experimenters sent a man into her room without any warning, who shouted at her threateningly. Not surprisingly, she was scared. At the same time her dog's heartbeat accelerated violently.[27]

Probably few dog owners would consider taking part in experiments such as these, but it would be relatively easy to repeat this kind of experiment using isolated rooms and non-frightening stimuli. For example, in an experiment with two dogs one could be fed and the other observed to see if it showed any signs of excitement at the same time, as in Blake's experiments with horses.

Although the experiments with rabbits involving Russian submarines may just be rumors, some properly controlled experiments with rabbits have recently been carried out in France, with much the same results. In these tests rabbits were monitored for stress by measuring the blood flow through their ears. This was done painlessly by placing a small clip over a shaved part of one ear, on one side of which is a miniature light source, and on the other side a photoelectric cell. In this way the amount of light that shone through the ear could be measured continuously. When rabbits feel stress, the blood vessels in their ears contract, the blood flow decreases, and more light passes through.

These experiments, conducted by René Peoc'h, involved pairs of rabbits taken from the same litter that had lived together in the same cages for months. They were compared with other pairs of rabbits that had been kept in isolation from each other in individual cages.

At the time of the experiment, each rabbit was placed in a soundproof cage, which also isolated it from electromagnetic influences. During each experiment the stress experienced by both rabbits was measured by monitoring the blood flow through the ears. Peoc'h found that when one of the rabbits experienced a stress, the other tended to experience a stress within three seconds. By contrast, the control pairs of rabbits that did not know each other did not show the same kind of telepathic connection. The differences between the pairs of rabbits that knew each other and the control pairs were highly significant statistically.[28]

It would be surprising if rabbits and dogs could influence one another telepathically in experiments but not in real-life situations. And indeed several people who own two or more dogs have told me they have noticed that the dogs seem to influence one another at a distance. For example, Margaret Simpson, of Castle Douglas in Scotland, has a Whippet and a Labrador. When they are out walking, the Whippet usually stays close to her while the Labrador ranges quite far and seems able to "call" the Whippet, especially when she finds a deer. "For no sensory reason that I could discern, the Whippet would get some sort of message and she would be off. It was exactly as if a thought had been transferred."

Some dogs also react when another dog to which they are bonded has had an accident or died in a distant place. A sheepdog in France, for instance, showed signs of great distress when her mother was killed on Réunion, over 6,000 miles away (page 115). Another example, reported by Major Patrick Pirie, concerned a female Labrador and her daughter. When the daughter was about nine months old and living with Major Pirie in Somerset, "for no reason at all and for the only time in her life she refused all food and remained quiet throughout the day. That evening we received a telephone call to tell us that her mother had been knocked down by a car and killed. I am convinced that she had some sort of perception and knew what had happened 100 miles away."

Another example concerns some Bernese Mountain Dogs. Josephine Woods wrote that "One of my dogs was diagnosed as having cancer and was at Cambridge Vet Center. Suddenly, just gone twelve midday, the other dog started howling and was very distressed for quite a while." Later that afternoon, the vet from Cambridge rang to say that the sick dog had been put to sleep at midday.

The examples of dogs, horses, and other species discussed in this chapter suggest that telepathy may be widespread within the animal kingdom.

## The common features of animal telepathy

Several characteristics of animal telepathy recur in very different species. These point toward the following conclusions about the basic principles of telepathy within species:

1. Animal telepathy involves the influence of animals on other animals independently of the known senses.
2. Telepathy usually occurs between closely related animals that are part of the same social group—in other words, it takes place between animals that are bonded with each other.
3. In schools, flocks, herds, packs, and other social groups, telepathic communication may play an important role in the coordination of the activity of the group as a whole.
4. At least in birds and mammals, telepathy has to do with emotions, needs, and intentions. Feelings communicated telepathically include fear, alarm, excitement, calls for help, calls to go to a particular place, anticipation of arrivals or departures, and distress and dying.

In the case of domesticated animals, these same principles apply to telepathic communication between people and animals that are bonded.

These common features of animal telepathy seem to pertain to much of human telepathy as well, especially to the most dramatic cases of human telepathy concerning distant deaths or accidents.

One of the most important conclusions of the investigations described in this book is that telepathy is not specifically human. It is a natural faculty, part of our animal nature.

## Does telepathy only work at a distance?

The fact that telepathic communication can occur when animals and people are *not* in sensory contact does not prove that telepathy still occurs when they *are* in sensory contact. Perhaps telepathy is a faculty that is switched on only when needed, as people switch on a radio intercom system when they are apart and turn it off when they are together.

On the other hand, psychic links or emotional bonds connect animals and people both when they are together and when they are apart. Telepathic communication may well occur when communication is also taking place through the known senses.

We do not assume that animals cease to smell someone when they see or hear them. We are happy to accept that the senses are not mutually exclusive and generally work together. I think the same goes for the invisible communication that occurs through psychic bonds: it normally works together with the senses. The psychic link is not switched off when they are together and switched on when they are apart; it is potentially there all the time.

The scientific study of animal telepathy is still in its infancy. As research on this subject progresses, I expect that telepathy will seem increasingly normal rather than paranormal, or beyond the normal. It is an aspect of the biology of social groups and social communication. It enables members of the group to influence one another even when they are beyond the range of sensory communication, and it may be of considerable survival value. If so, the capacity for telepathic communication must be subject to natural selection. Telepathy must have evolved. Its roots may lie deep down in evolutionary history, among the earliest social animals.

Part V

# The Sense of Direction

*Chapter 10*

# Incredible Journeys

Animals bond not only to members of their social group but also to particular places. Many kinds of animals, both wild and domesticated, can find their way home from unfamiliar locations. This attachment to places depends on morphic fields, which underlie the sense of direction that enables animals to find their way home over unfamiliar terrain.

The sense of direction also plays a vital role in migration. Some species, like swallows, salmon, and sea turtles, migrate from breeding grounds to feeding grounds and back again over thousands of miles. Their ability to navigate is one of the great unsolved mysteries of biology, as I discuss in the next chapter. Here too I think that morphic fields, and the ancestral memory inherent in them, could help provide an explanation.

## Homing dogs, cats, and horses

There are many stories of domesticated animals coming home after they have been left or lost far away. Some have achieved almost legendary status, like a collie called Bobby, lost in Indiana, who turned up at his home in Oregon the following year, having covered a distance of more than 2,000 miles. Such cases form the basis of Burnford's well-known animal adventure story *The Incredible Journey,*[1] made into a film by Walt

Disney, in which a Siamese cat, an old Bull Terrier, and a young Labrador find their way back home over 250 miles of wild country in northern Ontario.

Real-life incredible journeys happen over and over again, and remarkable cases are often reported in newspapers. On September 9, 1995, the London *Times* carried the following story:

> A sheepdog who was abandoned by car thieves has been reunited with his owner after walking 60 miles home. Blake, a ten-year-old Border Collie, was stolen with his kennel mate, four-year-old Roy, while they were in the back of Tony Balderstone's Land Rover. The thieves, who took the vehicle from Cley, Norfolk, dumped the dogs at Downham Market, 60 miles from Mr. Balderstone's home at Holt. Roy was caught in Downham Market two days later and returned to his master but Blake set off alone. Mr. Balderstone, a shepherd, said yesterday: "I knew he would make it home, as long as he did not get hit in a traffic accident or shot for worrying stock. I phoned farmers and gamekeepers along the route to alert them." Blake took five days to make the journey to Letheringsett, a mile from Mr. Balderstone's smallholding, where villagers recognized him.[2]

For every case like this that is reported in the newspapers, dozens must remain unpublicized. On our database we have ninety-five unpublished stories of homing dogs and sixty-one of homing cats. Some of these animals were abandoned or lost when they were away from home, but most were taken to live in a new home and later found their way back to their old one.

Nearly all of these animals were transported to the new place rather than walking there by themselves. They would therefore have been unable to note the smells, landmarks, or other details of the route. Most of the outward journeys were by car, but some were by bus or train, and one was by boat along Lake Zurich. Some animals were taken by indirect routes, but most of those who were spotted on their journey back were heading straight home, not following the route by which they had been taken. In any case, a dog or cat that tried to follow the roads or railway lines along which it had been carried on the outward journey would soon be squashed. Somehow the animals knew in what direction their home lay, even when they were in a place they had never been to before and to which they had been taken by an indirect route.

The clearest evidence that the animals' sense of direction does not depend on memorizing smells along the route, or other details of the outward journey, comes from cases where the animal was transported by airplane. During the Vietnam War, scout dogs used by U.S. troops were taken by helicopter to the war zones. One such dog, Troubles, was airlifted with his handler, William Richardson, into the jungle to support a patrol ten miles away. Richardson was wounded by enemy fire, and was airlifted to a hospital; the other members of the patrol simply abandoned the dog. Three weeks later Troubles was found back at his home at the First Air Cavalry Division Headquarters in An Khe. Tired and emaciated, he would not let anyone near him. He searched the tents until he found Richardson's belongings, then curled up and went to sleep.[3]

Another dog that could not have homed by following familiar smells along the route is Todd, a Labrador who fell overboard from his owner's yacht a mile off the coast of the Isle of Wight, in the English Channel. He was given up for dead by his owner after a four-hour search. But Todd was swimming across the Solent, the strait separating the Isle of Wight from the mainland. He swam at least ten miles, first across the choppy sea and then up the Beaulieu River, landing near his home. He had not attempted to reach the nearest land, on the Isle of Wight, a mile from where he fell into the sea, but instead headed homeward.[4]

Although most pet owners are astonished by the unsuspected homing powers of their animals, shepherds and other owners of working dogs are often well aware of this capacity. It is significant that Blake's owner, so confident of his return, was a shepherd. In the days when cattle were herded down from the Scottish Highlands into England, the drovers would send their dogs home on their own after they had delivered the cattle; the men stayed to work in the harvest. The dogs usually retraced the route they had taken on the way south, stopping at farms or inns where they had previously rested. The innkeepers fed them and were paid by the dogs' owners when they stopped the next year.[5] Before the Second World War, Lincolnshire farmers used to drive their animals in 20-mile stages to markets more than 100 miles away. When the animals had been sold, the drivers would turn their dogs loose to find their own way home to save paying their rail fare.

Some horses also find their way home across miles of unfamiliar country, and their homing abilities would probably be expressed much more frequently if they were not confined in fields and enclosures when taken to new places. The unwanted homing of horses is a nuisance, but sometimes a horse's ability to find its way back can be very useful.

One leisurely day, Jean Welsh was riding her horse in the Yorkshire countryside when she decided to explore an area that neither she nor the horse had ever been to before. After a while she realized she was lost. "I have a dreadful sense of direction and was a bit panicky. I dropped the reins on the mare's neck and said, 'It's up to you now—get us home!'" The horse carried on purposefully, stopping at a gate they had never seen before. So confident did she seem that Jean opened it. "Without any direction from me she continued on her way and appeared very much in control." They followed unfamiliar tracks until they eventually came to a place Jean recognized, much to her relief, not far from home.

## Other homing animals

The ability to home is widespread. In addition to stories about dogs, cats, and horses, we have on our database a tale of a flock of sheep that escaped from a farmer's field and traveled 8 miles to their native pasture, a story of a pet pig that homed from 7 miles away, and several stories of homing birds. One of the most vivid is about Donald and Dora, Easter ducklings raised by the Erickson family in Minnesota. Leni Erickson told this story:

> We built a fine pen in the backyard of our home in the inner part of Minneapolis. We fed [the ducklings] and gave them baths in a big plastic pool. They became quite the focus of our summer. Months passed and they were full size. What would we do when winter came? Finally in mid-August we decided to take them over to a pond in a large undeveloped park about two miles away. Mom said it was best if they joined their own kind and learned how to be wild before the snows came. We reluctantly agreed and let them go. Dad had marked their wings with paint so we could watch them mix around with the

wild ducks. We returned home sadly. Suddenly we heard neighbors out in the streets yelling and laughing. We ran out into the front yard and much to our amazement there on the top of the hill in the center of the street waddled Donald and Dora, quack, quack, quacking. They had found their way back home through woods and busy city streets.

In this case, the distance was modest, but some pet birds have found their way home over hundreds of miles. One was a magpie that had been adopted by the children of the Beauzetier family in Drancy, near Paris, after it fell out of its nest as a baby. For the summer vacation of 1995 the children went to stay with their grandparents near Bordeaux, taking the bird with them. While they were there the magpie escaped. The children were upset, and at the end of the vacation they had to return home without her. Soon afterward they saw her in a tree near their house. When they called her, she answered, and much to their delight came back to live with them. She had flown more than 300 miles.

Even more spectacular was the return of a pigeon belonging to Ken Clark, of Bakersfield, California. He gave the bird to some cousins who were visiting from Connecticut. He provided some feed and a cage for them to carry it in, and off they went. "One month later the bird was back! Its tail feathers were mostly gone. It was dirty and a real mess," Clark reported. His cousins had taken the bird all the way home, 3,000 miles away, but it escaped when they were trying to transfer it to a bigger cage.

The homing abilities of pigeons are no surprise, but they are by no means unique, and are shared by many other species.

## The bear problem

In North America, bears often cause problems in inhabited areas, raiding garbage containers, breaking into garages, and posing actual or perceived threats to human safety. Every year in British Columbia alone about 1,000 black bears and 50 grizzlies are shot, and in many other parts of the United States and Canada "a problem bear is a dead bear." In national parks and other places where killing the bears is unpopular, wildlife managers routinely capture the bears and move them to sparsely inhabited areas far away. The problem is that many of them return, even when they are taken 150 miles by helicopter.

In a study by the Alaska Department of Fish and Game, brown bears were captured in the remote Copper River Delta and fitted with radio collars so that their movement would be monitored. They were kept in metal cages and flown to the Ice Bay area, about 100 miles away, and released as family groups, compatible pairs, or alone. Of the thirteen radio-collared bears, ten started moving back in a homeward direction. Four of them either shed their radio collars or had defective collars and were lost, but the other six were traced back to their original home range in the Copper River Delta. The fastest moved at an average speed of 6 miles a day over the wild terrain.

The bears that did not move far from the release site were all young males, at an age when they were still establishing their home ranges. However, they were all shot by hunters.

The conclusion from this study was that moving problem bears has a high risk of failure. They "have a strong homing instinct and are capable of traveling relatively long distances in short periods of time."[6]

## Crocodiles

The same is true of problem crocodiles. In Australia the long established practice of capturing and removing the most aggressive crocodiles from outback swimming holes and flying them hundreds of miles away by helicopter turned out to be useless. The crocodiles found their way back, swimming between 6 and 19 miles a day. In a recent study, biologists from the University of Queensland tracked crocodiles by satellite so their movements could be observed in detail. Large male animals were transported one by one to remote locations in a net underneath the helicopter and released in the ocean near the coast. The crocodiles stayed near the point of release for at least two weeks, but then they set off "on an apparently purposeful and direct travel homeward." For example, one twelve-foot male swam around the northern tip of Australia, taking the most direct coastal route and traveling more than 250 miles altogether, moving up to 20 miles a day. Once back in his home river, he stayed there.[7]

The homing behavior of bears and crocodiles taken by helicopter to remote, unfamiliar places makes it very clear that these animals have a sense of direction that cannot be explained in terms of tracing their route home by smells, or by visual landmarks, or by any other obvious sensory means.

## Experiments with homing cats and dogs

Most pet owners are understandably reluctant to abandon their animals in unfamiliar places so as to study their homing behavior. Apart from my research with the dog Pepsi, described later, I know of only two series of experiments of this type.

The first were carried out in Cleveland, Ohio, over eighty-five years ago by the zoologist F. H. Herrick, with his own cat. His research began unintentionally when he took the cat in a bag from his home to his office at Western Reserve University, 5 miles away, traveling by streetcar. But when he let the cat out of the bag, it escaped and was home the same night. Puzzled by this direction-finding ability, he investigated it further by taking the cat in a closed container and releasing it at distances from one to three miles from his home. He established that the cat could home under a variety of conditions and from any point of the compass.[8]

The second series of experiments was carried out in Germany in 1931–32 by the naturalist Bastian Schmidt, who studied three sheepdogs. In each experiment, a dog was taken in a closed van by a roundabout route to a place it had never been to before. It was then released. Its behavior was observed and recorded by a series of trained observers stationed along its probable homeward route. It was also followed by silent cyclists, who were instructed not to communicate with it in any way.[9]

The first experiments were in the Bavarian countryside with a farm dog called Max. When Max was released for the first time in an unfamiliar place, he scanned the landscape in various directions, as if taking his bearings. After several trials he began to concentrate on the direction of his home, looking resolutely homeward, and after half an hour he set off. He avoided going through woods, hid from passing cars, and circumvented farmhouses and villages. After traveling for just more than an hour, he came out on the familiar road into his village and galloped home. The distance he covered was about 6 miles.

In the second trial he was released in the same place and, after hesitating for only five minutes, set off along the same route as before, but this time he took a shortcut and reached home in forty-three minutes. In a third trial he took longer, because he was forced by some heavy traffic to take a long detour.

From the observations of Max's behavior, Schmidt concluded that he "made *no* use of the sense of smell, although this sense is so important to a dog." He did not sniff at the trees or at the ground, nor did he try to pick up any trail. There was no reason for him to do so, Schmidt concluded: "The picking up of a trail, human or canine, could mean nothing for a dog trying to return to his home."[10] Nor could he have used his sight to determine the homeward direction, because he could not see any familiar landmarks.

Schmidt then did some experiments with a city dog named Nora. She lived in Munich and, for the purpose of the experiment, was taken early in the morning to a part of the city she had never been to before, more than three miles from her home. When she stepped out of the carrying basket, she found herself in a great square—Johannisplatz in Bogenhausen; she lived close to the Tierpark. When she was first released, she behaved very much as Max had done; she spent about twenty-five minutes getting her bearings, looking principally in the direction of her home, and then she trotted off in the right direction. All went well until she encountered a frolicsome dog in the Tassiloplatz who led her astray. After some time she took her bearings again and once more set off in a direct line toward her home. The journey took ninety-three minutes, including the time spent taking her bearings, playing, and straying.

For the second experiment, nearly six weeks later, Nora was released in the same place as before. This time, again like Max, she took only five minutes to get her bearings, and set off along the route she had followed before, as far as the Tassiloplatz. This time there was no distraction, and she ran straight home, arriving thirty-seven minutes after she was released.

Like Max, Nora was not sniffing and seemed to be paying no attention to smells. She could not see any familiar sights, since there were many streets of houses between the release point and her home. Since neither smell nor sight seemed able to explain her behavior or that of Max, Schmidt concluded: "Here we are confronted with an enigma, the mystery of an unknown sense, which one might perhaps describe simply as the sense of orientation."[11]

Schmidt then tried out three similar experiments with another country dog, but they were all failures. The dog always went off in the wrong direction. This is a salutary reminder that dogs, like people, differ in their abilities; some have a better sense of direction than others.

Elizabeth Marshall Thomas, in her engaging book *The Hidden Life of Dogs,* has recorded her observations of dogs left to their own devices. She came to similar conclusions. One of her dogs, a Husky called Misha, had excellent navigational skills and went on journeys that took him up to 20 miles away from home. Misha's mate, Maria, did not get lost when she accompanied him so long as she followed his lead. But she almost always got lost when she went out on her own. Then she used her own way of getting back: she simply sat forlornly on someone's doorstep. Sooner or later, the homeowner looked on her collar for her telephone number and called Thomas, who collected Maria in a car.

The first question Thomas asked when she started studying her dogs was about the nature of Misha's navigational skills. "But this question, I was never able to answer."[12] Misha did not seem to be navigating by landmarks, since once he had reached a destination he might easily take another route home. Did he use the stars or the position of the sun? The sound of the nearby Atlantic Ocean? Odors floating in the air? Thomas said, "I didn't know, and could learn nothing by watching his sure trot, his confident demeanor."[13]

## Multiple destinations: Experiments with Pepsi

As we have seen, many animals can return home from strange places. But some are able to find places other than their own home across unfamiliar terrain. They seem to have a sense of direction for *several* places.

The most remarkable direction-finding dog I have met is Pepsi, a Border Collie–terrier cross who lived in Leicester with her owner, Clive Rudkin. By 1995, on fourteen occasions Pepsi had made journeys all over the city of Leicester after escaping from Clive's own house or from his relatives' houses. Each time she arrived within hours at the house of a friend or another relative. Most of these journeys covered at least three miles in several directions. Altogether, Pepsi found her way to six different destinations. She had previously been taken to all six places by car, but she had never walked to any of them, and during the car journeys she was usually on the floor and unable to see out the windows. On one occasion Pepsi escaped from Clive's parents' house, 4 miles northeast of his home, and turned up at a friend's house 5 miles north of there. She had been driven to this house directly from Clive's own house but never from his parents' house.

Apart from these adventures, spaced over a period of four years, Pepsi was never left to roam the streets on her own; she was always accompanied when walking. She had never gotten lost or suffered any injury, and Clive felt sufficiently confident of her abilities to agree to doing two experiments in which she was left to find her own way home from unfamiliar places. These experiments were filmed for BBC, and a report was televised in 1996.[14]

In the first test, Pepsi was released in a park 2 miles southwest of her home, and followed by a BBC cameraman. She made her way home by a slightly indirect but scenic route, following the bank of a river for much of the way. The problem with this experiment was that she soon noticed the cameraman following her, and for some of the journey started following him. Since she insisted on interacting with him, he was unable to act as a detached observer, and it is hard to know how much Pepsi's journey was influenced by his presence.

For the second experiment, Pepsi was equipped with a Global Positioning System (GPS) monitor in a pouch attached to her back. This device, about the size of a portable telephone, accurately recorded her position within about 10 yards by means of signals from satellites. Our plan was to leave Pepsi on her own in an unfamiliar place very early in the morning, to minimize the danger from traffic, and to track her movements by satellite, her positions being recorded automatically at one-minute intervals.

On the summer solstice of 1996, as latter-day Druids were celebrating the sunrise at ancient megaliths, Clive and I were in Ethel Road, Leicester, leaving Pepsi on a street corner 2 miles east of Clive's house. She had never been taken to this place before. We went there in a taxi; she traveled on the floor and was unable to look out of the windows. She looked at us quizzically as she sat on the curb and watched us disappear in the same taxi. We had taken the precaution of putting a message on her back explaining to anyone who found her that she was taking part in an experiment, and we had informed the police in case she strayed.

We went back to Clive's house and waited. We had left her at 4:55 A.M., and we expected her to turn up at Clive's house, or possibly at his parents', within two hours at most. By 9:00 A.M. she had still not arrived, nor had she turned up at Clive's parents' house, and we were getting very worried. Finally, Clive thought of checking the house of his sister, who was away on vacation. And there we found Pepsi, lying calmly on the grass in the back garden. Pepsi had not been taken to this house for at least six months, and had never made her own way there. She had, however, escaped from this house on two occasions in previous years and gone to a friend's house some 4 miles to the southwest.

In retrospect, we could see that this was the best choice for Pepsi, because it was the nearest house she knew, only a mile east of the place where we had left her. The record on the GPS device told us that Pepsi had first gone about 500 yards north—the direction opposite to the one from which we had departed in the taxi. She had then spent at least eight minutes going backward and forward in the surrounding streets, as if getting her bearings. She next headed due east for three-quarters of a mile to the neighborhood of Leicester General Hospital and spent seven minutes roaming around the hospital buildings. Then she went straight to Clive's sister's house, about 500 yards south of there (Figure 10.1).

Pepsi could not have found the house by scenting it, because that morning there was a steady northwest wind, and at no point in her journey was she downwind of the house or its neighborhood.

After this experiment, Pepsi escaped on four more occasions and

*Figure 10.1 A map of part of Leicester showing where Pepsi was released (A), the succession of places she visited, as revealed by the Global Positioning*

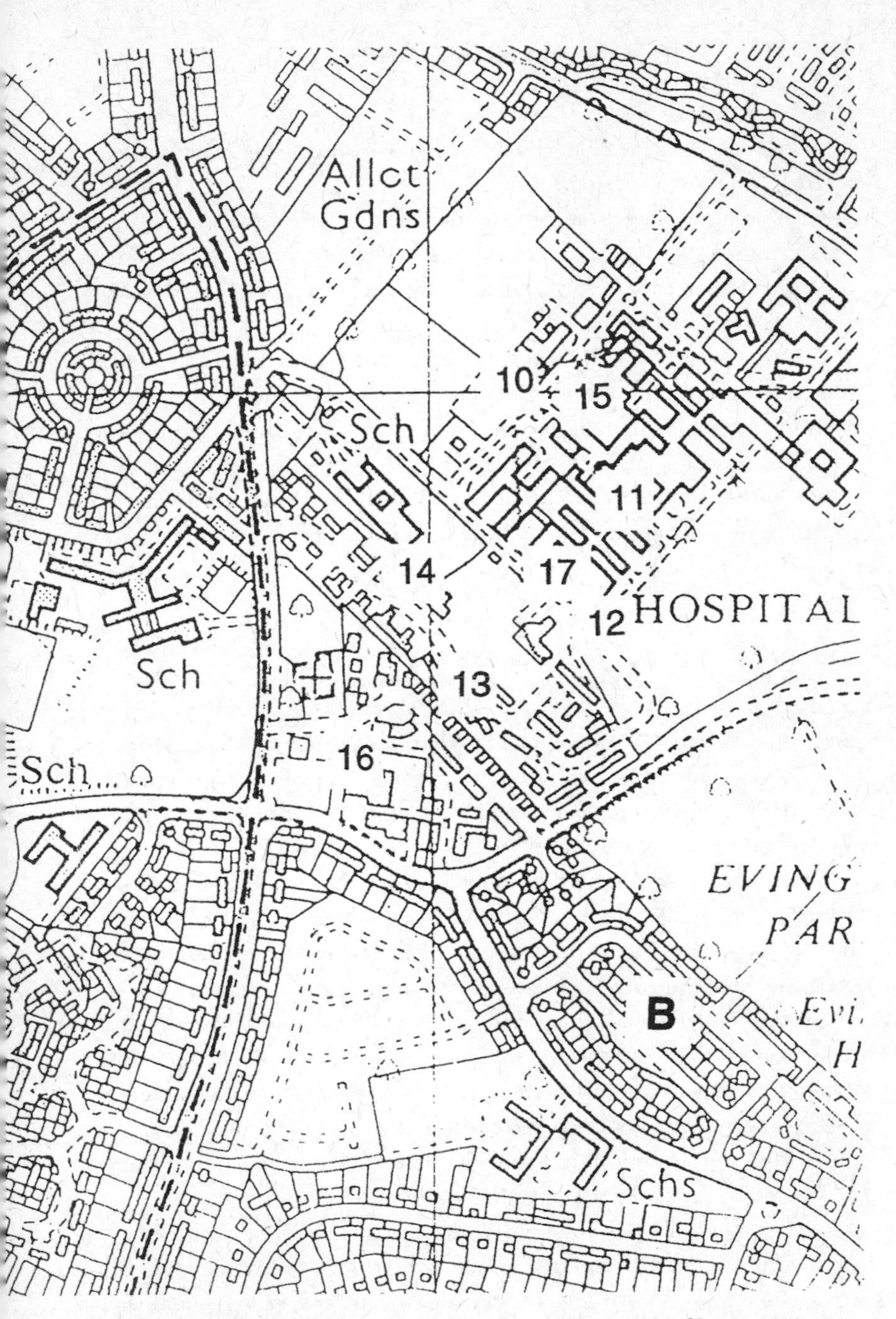


*System monitor she carried on her back, and the house of her owner's sister (B), where she ended up.*

made more journeys across Leicester to houses she knew, and also to the home of Clive's brother, which she had never made her own way to before.

Pepsi's sense of direction somehow enabled her to know where she was in relation to several houses, and to know where those houses were in relation to one another, even though she had been taken from one to the other by car, not looking out of the window.

One way of thinking about this would be to suppose she had a kind of mental map. But this is too abstract and too anthropomorphic a metaphor. And even a mental map would not enable her to know where she was when abandoned in an unknown place. Maps are useful if you know where you are and where you want to go. But if you do not know where you are, a map is not much help. Instead of a map, Pepsi seems to have a sense of direction.

## The sense of direction

How could a sense of direction work? Whatever the physical basis, I suppose that the animal somehow *feels* that a familiar place is in a certain direction, perhaps through some kind of "pull" toward it. And it may also feel its nearness or distance.

In the simplest case, that of homing behavior, the animal feels pulled toward its home, and if it drifts off course, as Nora did in Munich when she met the playful dog, it can take its bearings again and reorient itself toward its home (Figure 10.2A). One metaphor would be magnetic attraction. Another would be a connection to its home by an invisible elastic band. Either way, there would be a kind of pull in the homeward direction, and a feeling of "getting warm" on approaching the home. This feeling of getting warm is also consistent with the ability of animals transported in cars and other vehicles to know when they are nearing their home, as I shall discuss in Chapter 12.

I suggest that this pull toward home takes place within a field connecting the animal with its environment. The animal builds up a familiarity with its home environment. Its field of activity within its familiar environment involves the building up of memories.

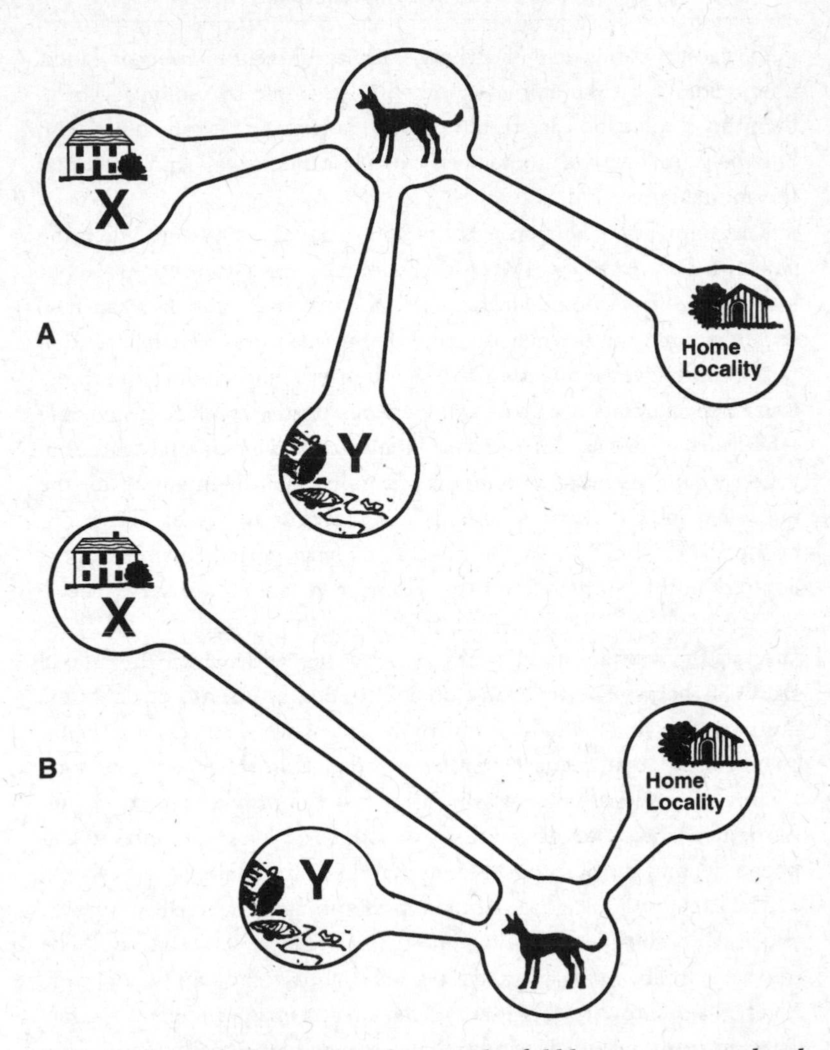


*Figure 10.2 A: Connections through a morphic field between an animal and its home and other places of importance to it. B: The animal is in a different location, and hence the connections to its home and other places give it different directional information.*

I suggest that this field of activity, with its inherent memory, is a morphic field. And if the animal is linked to its home by a morphic field, then this connection can stretch like an elastic band. It can continue to link the animal with its home even when it is miles away. And it can pull the animal homeward.

The animal's connection with its home may remain latent when the animal is busy foraging or exploring. Animals away from home are not being pulled back home all the time they are away, but they can usually find their home when the time comes to return. Their intention to go home gives motivation to their behavior, but finding the direction to their home or other place depends on the connections already established with particular places. Homing behavior, like navigation in general, depends on a combination of motivation or intention on the one hand, and connections to significant places on the other. These connections were built up in the past, and I suggest that this memory is inherent in the morphic fields that connect the animals to these places.

The morphic field that connects an animal to its home is closely related to the morphic field of the social group with which the animal shares its home. The first kind of field underlies a sense of direction; the second, telepathic communication. But as we shall see in Chapter 13, the fields that connect animals to other animals not only provide channels for telepathic communication but can provide directional information as well. We have already encountered some examples of this phenomenon when cats or other animals in distress called their owners.

The analogy of the elastic band in itself implies directionality. Imagine you are blindfolded and holding one end of a long, taut elastic band. The other end, many yards away, is attached to a place or held by a person. You feel a pull toward that place or person and in a particular direction. This attraction, or pull, toward a goal can be modeled mathematically in terms of dynamical attractors within morphic fields.[15]

When an animal has several familiar places to which it can travel, it presumably experiences pulls or attractions to different places. For each place it may also have a feeling of nearness or distance. If the animal is in a different position, then the pulls are oriented in different directions (Figure 10.2B).

In the light of these ideas, I imagine that when Pepsi was left in an unfamiliar place, as she took her bearings, she sensed the direction of various familiar houses, including her own home, and Clive's parents' and sister's houses. She may also have felt that the sister's house was the closest, and hence set off in that direction.

The elastic band provides one metaphor for these pulls. Magnetic attraction provides another. One of the advantages of the magnetic metaphor is that it raises the possibility not only of an attraction toward particular places but also of repulsion from them. It is possible that locations that animals fear—for example, places where they had traumatic experiences—may repel rather than attract them, even at a distance. They may also have the feeling of "getting warm"—or perhaps "getting cold"—as they approach such places, and this feeling may be one of fear rather than excitement. This would explain the fearful reactions of some dogs as they are driven in cars toward the vet.

The direction-finding abilities of domesticated animals make good sense when we look at them in a larger biological and evolutionary context.

## Home ranges of wild and feral animals

Animals that have a home base—bees with their hive, robins with their nest, wolves with their den—are familiar with the surrounding area. The part of the familiar area they visit repeatedly is the *home range,* the size of which may vary from day to day and from season to season. And within the home range they may have a *territory,* an area they protect. A domestic cat, for example, may have a territory about 100 yards in diameter that it knows intimately and which it protects, but it may also have a larger home range extending up to a mile or more from its home.

We too have our familiar areas, and within them a home range covering our immediate neighborhood, places we shop, work, and play, places where we visit family and friends, areas in which we walk our dogs, and so on. Within these home ranges are the territories we defend, usually our homes and gardens.

Generally speaking, in finding our way around our home range we depend on landmarks and familiar features, and the same is true of animals. Familiar sights, sounds, and smells enable animals to know where they are and to find their way to familiar destinations. They do not need a special sense of direction when they are moving within the home range and following familiar routes.

But of course an animal's familiar area has not always been familiar to it. Every young animal has to get to know it for the first time, even if it is already well known to older members of the group. Each individual animal, when it settles in a new place, needs to explore its surroundings. When it is exploring new terrain, it cannot rely on memory to find its way home, unless it retraces its steps.

Animals that have been out exploring may indeed find their way back by following landmarks or a scent trail. But another way to do so is by *biological navigation*—"the capacity to orient toward a goal, regardless of its direction, without reference to landmarks."[16] This way an animal can find its way back by taking shortcuts and without needing to memorize all the features of the outward journey. Navigation also enables it to find its way home from an unfamiliar place when it has not had an opportunity to learn the features of the outward route. If an animal is being chased by a predator, for example, it may escape by running into an unknown place without remembering all the details of its path. Likewise, hunting animals often move away from the familiar pathways of their home range as they follow their prey. Birds may be blown off course by strong winds, and aquatic animals can be carried into unfamiliar waters by currents. In all these circumstances, the animals need to be able to navigate to find their way back home.

As a general rule, the larger the home range, the more important navigational skills will be for getting home or finding other important places within the range. Some packs of wolves, for example, possess and defend enormous territories. In northern Minnesota, where deer are relatively plentiful, wolf territories are 50 to 100 square miles. In Alaska, where wolves mainly prey on moose, a territory may be 800 square miles. On Arctic islands where prey populations are sparse, a pack's territory may cover thousands of square miles. One pack on Ellesmere Island, northwest of Greenland, was observed to range more than 5,000 square miles in a six-week period.[17]

Considering that wolves, the wild ancestors of dogs, possess navigational skills that enable them to find their way around in such vast areas, the homing abilities of domestic dogs seem less astonishing.

Feral dogs have smaller home ranges than wolves, but they still display an impressive ability to find their way around. In central and southern Italy, for example, packs of free-ranging dogs are common, and several have been studied by radiotracking and by visual observation. In one of these studies, in a mountainous part of the Abruzzo region, the overall home range of one pack was 22 square miles, and within this range, core areas were frequented much more than others, especially near the den and the rubbish dumps where the dogs found food. The home range shifted from season to season and from year to year, and the pack established new core areas as new food sources were found.

On several occasions dogs went on major excursions outside the home range, seemingly to explore. As a result of one of these expeditions, a bitch established a new den 10 miles away. Interestingly, the home range of these feral dogs was between the territories of two wolf packs, with some overlap. The wolves' ranges were considerably larger than the dogs'. One of them covered 110 square miles.[18]

Cats generally have much smaller home ranges, though some farm cats may cover up to 50 acres (under a tenth of a square mile).[19] Males generally have larger ranges than females, and some male feral cats in the Australian bush range more than two square miles.[20]

When animals such as cats, dogs, and wolves find their own way around, it is hard to know how much of their navigational skill depends on keeping track of the route they have followed, using memory and their normal senses, and how much it depends on a more mysterious sense of direction. Maybe these factors all work together most of the time.

Although domestic dogs and cats normally have less freedom to move around on their own than do their wild and feral relatives, they can tell us more about this sense of direction precisely because they are less free. To study the sense of direction in free-living animals, it is necessary to capture them and then take them to an unfamiliar place before releasing them. Relatively few studies of this kind have been carried out. But large numbers of dogs and cats are moved around in cars and other vehicles. The fact that they are transported passively and often go to sleep on the journey means they cannot study and remember the details of their route. Yet, as we shall see in Chapter 12, they often know when they are nearing their destination.

## Homing pigeons

The most impressive homing performances are by birds, and the best studied are pigeons. The distance records, however, are held not by pigeons but by several wild species. Adélie penguins, Leach's petrels, Manx shearwaters, albatrosses, storks, terns, swallows, and starlings have all been known to home from more than a thousand miles away.[21] When two Laysian albatrosses were taken from Midway Island in the central Pacific and released from Washington State, 3,200 miles away, one returned in ten days, the other in twelve. A third came back from the Philippines, more than 4,000 miles away, in just over a month.[22] In an experiment with Manx shearwaters, birds were kidnapped from their nesting burrows on the island of Skokholm, off the coast of Wales. One was released in Venice, Italy, and was back within fourteen days. Another returned in twelve and a half days from Boston, Massachusetts, a journey across the Atlantic of 3,000 miles.[23]

Although their range is more limited than that of such seabirds, racing pigeons are the obvious choice for detailed research. They have been bred and selected for their homing ability over many generations. Racing pigeons can fly home in a single day from a place they have never been to before, hundreds of miles away. The techniques for keeping and training them are well known. And the birds are relatively inexpensive.

Numerous experiments on homing have already been carried out with pigeons. Nevertheless, after more than a century of dedicated but frustrating research, no one knows how they do it. All attempts to explain their navigational ability in terms of known senses and physical forces have proved unsuccessful. The best-informed researchers in this field readily admit the problem: "The amazing flexibility of homing and migrating birds has been a puzzle for years. Remove cue after cue, and yet animals still retain some backup strategy for establishing flight direction,"[24] wrote one scientist. Another says, "The problem of navigation remains essentially unsolved."[25]

To appreciate the difficulty of solving this problem, it is necessary to consider the various theories of pigeon navigation that have been put forward over the years and to see why all have proved inadequate.

The theory, first proposed by Charles Darwin, that the birds remember the twists and turns of the outward journey has been refuted by taking pigeons to an unfamiliar point of release in dark vans, within rotating containers, by devious routes. Some were even anesthetized throughout the journey. When they were released, they flew straight home.[26]

The theory that they rely on familiar landmarks has also been ruled out. Pigeons can find their way home by following landmarks in familiar territory, but they can also return from unfamiliar places hundreds of miles away where they cannot see any recognizable landmarks. And in experiments carried out in the 1970s, pigeons were even temporarily blinded by being fitted with frosted-glass contact lenses. They still found their way home over great distances, although they tended to crash into trees or wires very near their loft. They had to be able to see in order to land properly. But they found their way from many miles away to within a few hundred yards of the loft without being able to use their eyes.[27]

The sun navigation theory postulated that the pigeons use the position of the sun to work out their latitude and longitude, comparing its angle and motion at the point of release with those at home. This theory has been refuted in two ways. First, pigeons can home on overcast days and can even be trained to home at night. This means that being able to see the sun is not essential for homing. Secondly, navigating by the sun is possible only with the help of a very accurate clock. This was a lesson learned by humans in the eighteenth century through various attempts to determine longitude at sea, of great importance to naval navigation.[28] But when pigeons have their internal clock shifted by six or twelve hours (by keeping them in artificial light for part of the night and in darkness for part of the day), when they are released on sunny days, they are at first confused, and set off in the wrong direction. But they soon correct their course and fly home. On cloudy days they set off in the right direction straightaway. These results show that pigeons may use the sun as a kind of compass, but the sun is not essential for their knowing the direction of their home.[29]

The theory that pigeons smell their home from hundreds of miles away, even when the wind is blowing in the wrong direction, seems extremely implausible. Nevertheless, it has been tested in a variety of ways. In most of these experiments, the pigeons could still find their way home even if their nostrils were blocked up with wax, their olfactory nerves severed, or their olfactory mucosa anesthetized. They may use smell in familiar regions where they can recognize the odors carried on the wind, but their ability to home from unfamiliar places cannot be explained in terms of this sense.[30]

Finally, there is the magnetic theory. Could it be that the pigeons have a magnetic sense, a biological compass? Some birds can indeed detect the earth's magnetic field, and there are two ways in which they may do so: first by using mineral-iron magnetic receptors in the upper parts of their beaks, and second by "seeing" the magnetic field through the magnetic sensitivity of pigments in their eyes.[31] The problem is that even if pigeons do have a compass sense, it would not tell them where their home was. If you are taken to an unfamiliar place and given a compass, you know where north is, but not where your home is. The compass would be useful for keeping your bearings, but you would need to know where home was by some other means.

But what if the compass sense were so sensitive that it could give information on latitude? It could do this in two ways: first, by detecting the small changes in the earth's magnetic field strength at different latitudes and, second, by detecting the dip of the magnetic field. At the magnetic North Pole the needle points downward; at the equator it is horizontal, and in between, the angle varies according to the latitude. But to detect changes in latitude, the pigeons' magnetic sense would have to be very accurate indeed. In the northeastern United States, for example, over a distance of 100 miles in a north-south direction, the average field strength changes by less than 1 percent, and the angle of the field by less than one degree. Even if pigeons had such an accurate compass sense, it would give no information on longitude, on how far east or west they were from home. Pigeons can home from all points of the compass.

In any case the magnetic hypothesis has been tested directly by attaching magnets to pigeons. These should confuse their magnetic sense, if they have one, and yet birds with magnets attached to them get home as well as control birds with non-magnetic attachments of similar size and weight.[32]

The invalidity of all these theories leaves pigeon homing essentially unexplained. I myself think that pigeons' feats of navigation can be explained only in terms of a sense of direction, as I have already discussed in relation to the navigation of dogs, cats, and other animals. No doubt pigeons' navigation can be aided by using the sun's position and perhaps even a magnetic sense to help them keep their bearings so as to stay on course. But without the directional pull through the morphic field connecting them to their home, they would be lost.

## The human sense of direction

Our hunting and gathering ancestors were subject to the same selection pressures as other animals. Groups or individuals who traveled away from their home base and failed to find their way back probably perished, unless they were fortunate enough to find another human group that allowed them to join.

Until very recently, traditional peoples such as Australian Aborigines, the Bushmen of the Kalahari, and the navigators of Polynesia were famous for their sense of direction. Here were human beings whose abilities far exceeded ours. Laurens van der Post, traveling in the Kalahari with some Bushmen, after many miles of following a twisting trail had no idea where they were or in which direction their camp lay. But his companions had no doubt. "They were always centered," van der Post wrote. "They knew, without conscious effort, where their home was."[33]

One of the most spectacular demonstrations of this ability was given by Tupaia, a dispossessed high chief and navigator from Raiatea, near Tahiti. Captain James Cook met him in 1769 on his first great voyage of exploration, and invited him to travel on board the *Endeavour.* During a journey of more than 6,000 miles, via the Society Islands, around New Zealand, and along the Australian coast to Java, Tupaia was able to point toward Tahiti at any time, despite the distance involved and the ship's circuitous route between latitudes 48°S and 4°N.[34]

By contrast, civilized peoples, and especially modern urban people, have so many artificial aids to navigation, such as signposts, maps, and compasses—and now satellite global positioning systems—that a sense of direction is no longer essential to survival. It is neglected during our education, and very little attention has been paid to the subject by institutional science.

Nevertheless, the sense of direction has not atrophied altogether in modern people.[35] Most of us are vaguely aware of this sense, if only by comparison with other people who tend to get lost more easily or who are much better at finding their way. Nevertheless, in the absence of artificial aids, most modern people are poor navigators compared with many nonhuman animals. And this is no doubt why we find the abilities of dogs and cats so fascinating, and why homing pigeons are especially intriguing. They can do something we can't. They have sensitivities that we have lost.

*Chapter 11*

# Migrations and Memory

Homing and migration are closely related. Cycles of migration can be thought of as a double-homing system. For example, English swallows migrate 6,000 miles in the autumn to their winter feeding grounds in South Africa, crossing the Sahara Desert on the way. They return to their English breeding grounds in the spring, often to the very same place where they nested the previous year. They home to Africa, and then they home to England again.

More amazing still is the instinctive ability of young birds to home to their ancestral winter quarters without being guided by birds that have done it before. European cuckoos, raised by birds of other species, do not know their parents. In any case, the older cuckoos leave for southern Africa in July or August before the new generation is ready to go. About four weeks later the young cuckoos find their own way to their ancestral feeding grounds in Africa, unaided and unaccompanied.

The Icelandic black-tailed godwit achieves remarkable feats of navigation and synchronization. Pairs are monogamous, breeding with the same partner year after year in Iceland, where they arrive from mid-April to mid-May. Yet the partners do not winter together. Instead the sexes migrate separately at the end of the breeding season, with the females leaving first and the males staying behind to care for the young. Some of the birds migrate to England and others to Portugal. Typically members of each pair spend the winter 600 miles apart and fly back separately too. Yet they manage to return not only to the right place, where they meet their partner again, but at the right time as well. The partners usually arrive within three days of each other. They converge from very different starting points and coordinate their journeys to synchronize their arrival.[1] No one knows how they do it.

Even insects can migrate enormous distances to places they have never been before. The most famous is the monarch butterfly. Monarchs born near the Great Lakes, in the northeastern United States, travel some 2,000 miles south to winter by the millions on certain "butterfly trees" in the Mexican highlands. Then they migrate north in the spring. The migrants of this first generation die after breeding in the southern part of their range, from Texas to Florida. Their offspring continue the northward migration to the Great Lakes region and southern Canada, where they breed for several generations. In the autumn, the migrants of the new generation head south toward Mexico to winter on the ancestral trees; but these are three to five generations away from their ancestors that wintered in the same spot last year. This migratory cycle is continued through successive generations, and no individual butterfly experiences more than a part of it.[2]

How do these animals find their way to these ancestral destinations? Are they navigating toward a goal, as homing pigeons do, using a sense of direction? Or are they merely following a series of genetically programmed instructions that tell them to go in a particular direction, orienting by means of the sun, the stars, and a magnetic sense?

In this chapter I argue that the genetic programming theory is inadequate to account for most migratory behavior. Instead, I suggest that migrating animals often rely on a sense of direction that enables them to navigate toward their goal, to which they are connected through morphic fields. I propose that their migratory pathways involve an ancestral memory that is inherent in these fields.

But just as homing birds navigating toward a goal with a sense of their direction may also make use of a compass sense and the sun's position to help them keep their bearings, so migrating animals may also make use of magnetic and celestial clues.

## The sun, the stars, and the compass sense

Migratory birds are usually imagined by biologists to have an inborn program that directs the migratory process on the basis of compass orientations derived from the sun, the stars, and a magnetic sense. In the scientific literature this is called an inherited spatiotemporal vector-navigation program.[3] But this impressive-sounding technical term merely restates the problem instead of solving it.

The main evidence for the role of stars is that when migratory birds are kept in cages in a planetarium at the beginning of the migration season, they tend to hop in the appropriate direction of migration according to the rotating pattern of "stars." In the Northern Hemisphere, the point around which the stars rotate is the celestial North Pole, and hence the movement of the stars can serve as a kind of compass.

However, in the real world, migrants can still find their way in the daytime or when the sky is heavily overcast.[4] For example, in a radar tracking experiment based in Albany County, New York, it was found that uninterrupted overcast skies lasting several days did not disorient nocturnal migrant birds of various species. There were "not even subtle changes in flight behavior."[5] Thus a star compass does not seem to be essential for their orientation.

Then what about a magnetic compass sense? Some species do indeed seem to be sensitive to the earth's magnetic field,[6] and captive migratory birds kept in cages change the direction in which they hop if the magnetic field around them is changed.[7]

Although a compass sense and the rotation of the stars can help birds in their orientation, knowing the compass directions cannot tell them where they are and where their goal is.

First, the genetic programming theory proposes not that they know where they are going but that they fly in a programmed direction. There is a big difference between navigating to a goal and following a series of directions. For one thing, goal-directed navigation is more flexible. If you are trying to reach a city by road and you get lost, you can find your way there by a new route if you know where you are trying to go. If you do not know where your destination is, however, and you simply follow a series of directions, such as "drive 75 miles northeast and then 20 miles north," you will not be able to adapt to any emergencies that cause you to lose your way.

A programmed migratory pathway would have to be very finely adjusted if the animals are to find their way to the wintering area from different starting points and then return to the same places the next spring. Swallows from western Ireland, eastern England, and northern Germany, for example, set off in different directions and fly by different routes before they converge on the Strait of Gibraltar, where they cross over to Africa. On the return journey, they would have to be programmed to diverge at particular points after crossing back into Europe, and then to follow distinctive routes to their destinations. Such a rigid system would be inflexible, and any birds that were blown off course would have little chance of finding their way back to their breeding grounds.

Second, these hypothetical programs would have to be built up on the basis of chance mutations and natural selection over many generations. This would make it difficult for new migratory patterns to evolve, and it would prevent the animals from adapting rapidly to changing circumstances.

And third, the only remotely plausible mechanism for programmed migratory behavior is in terms of a magnetic sense, combined with information from the sun and stars. The problem is that not only does the earth's magnetic field vary throughout the day and with the seasons of the year, but the magnetic poles themselves wander as well. The north magnetic pole is not at the geographical North Pole; in 2005 it was in northern Canada, near Ellesmere Island (around 111°W and 81°N). In 2009 it was moving toward Russia at about 40 miles per year. This means that compass needles do not point to true north but rather deviate from it. The angle of deviation, called the declination, varies from place to place, being greatest in northern latitudes. Human navigators using compasses have to correct for this magnetic declination depending on their latitude and longitude, using correction factors that are continually updated as the magnetic poles wander. No animal could be genetically programmed to make such corrections.

In addition to the wandering of the poles, the general pattern of the earth's magnetic field itself changes considerably over the years, with quite large changes over timescales of a couple of centuries (Figure 11.1). Any

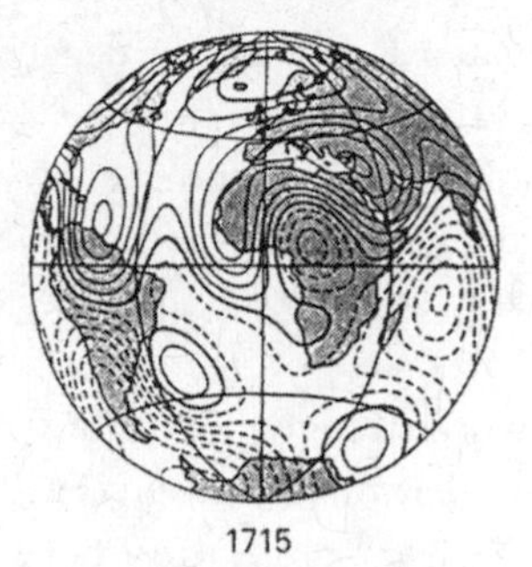

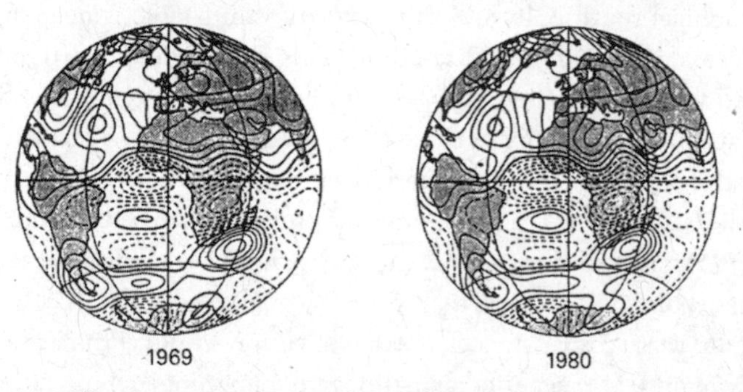


*Figure 11.1 The changing magnetic field of the earth over the last few centuries. The contours indicate the strength of the field at the boundary between the molten core and the mantle. The lines of force come out of the Southern Hemisphere and flow back into the Northern Hemisphere. The solid contours represent the intensity of magnetic flux into the core; the dotted lines represent flux out of the core (after Bloxham and Gubbins, 1985).*

genetically programmed magnetic navigational system that depended in detail on the earth's magnetic field would be disrupted by these changes. A genetically programmed system would have to be reprogrammed continually. But the timescales of the changes in the earth's magnetic field are too short for natural selection to be able to adjust the frequencies of the supposed "migration genes."

Natural selection would probably work strongly against any rigidly programmed system. We already know that dogs, cats, pigeons, bears, crocodiles, and other animals can home from places they have never been before. They show true goal-directed navigation. This behavior seems to depend on a link with their homes that gives them a sense of direction, enabling them to locate their home from wherever they are. The use of this more flexible sense of direction would probably be favored by natural selection over rigid genetic programming, even if such programming were possible.

Finally, any kind of programmed migration that depended on a magnetic sense would have to be extremely adaptable in periods of revolutionary change in the earth's magnetic field. At varying intervals, the magnetic poles reverse, so that the magnetic north pole is near the geographical south pole, and the magnetic south pole is near the geographical north. In the last 20 million years, the magnetic north pole has flipped to the south pole forty-one times, and forty-one times flipped back again.[8] (The history of these polar reversals has been reconstructed from the direction of magnetization in magnetic rocks, which provide a fossil record of the magnetic polarity prevailing at the time they were formed. A reversal of polarity is shown by the reversed magnetization of successive deposits of rock.)

Under these circumstances, natural selection would eliminate animals following a rigid magnetic-navigation program. Since all migratory animals today are the descendants of ancestors that have survived some 80 magnetic reversals in the last 20 million years, all must have had ancestors capable of reaching their goals in spite of reversals in the earth's magnetic polarity.

But what if animals can calibrate their magnetic compass sense on the basis of clues from the sky, such as the direction of the sunset and the rotation of stars around the north celestial pole? Research on migrating

Savannah sparrows in America has shown that their compass sense can indeed be calibrated through observation of the stars and can also be recalibrated throughout the lifetime of individual birds.[9] Species capable of such calibration could preserve a simple compass sense in spite of variations in the earth's magnetic field.

But although some migratory species may use the earth's magnetic field to help keep them on course, this is very different from a navigation system that can tell them where they are and where their goal is.

Even if animals could somehow inherit a mental map and know where their goal is, it is very unlikely that they could navigate simply on the basis of a compass sense and the observation of the sun and stars. After all, until the eighteenth century, not even the most sophisticated sailors could navigate with any accuracy on the basis of maps, compasses, and celestial observations. They used the sun's elevation at noon to determine their latitude, their north-south position. Magnetic compasses helped them keep their bearings. But they were incapable of working out their longitude, their east-west position. It was not until the invention of the chronometer by John Harrison, 250 years ago, that a precise determination of longitude became possible at sea, permitting accurate marine navigation.[10]

## Oceanic migrants

Fish, such as salmon and eels, can migrate over thousands of miles, and the movements of the sun and stars cannot explain their orientation: they could hardly observe the heavens with any precision from under the surface of the sea. They must have other means of finding their way. Smell probably plays an important part when they are near their destination, and in the case of salmon, there is good evidence that they "smell" their home river when they approach its estuary.[11] But smell cannot explain how they get near enough to the right stretch of coastline from oceanic feeding grounds hundreds or thousands of miles away. Similar problems arise in trying to understand the migrations of marine turtles.

Baby green turtles that have hatched on the beaches of Ascension

Island, in the middle of the Atlantic, find their way across the ocean to their ancestral feeding grounds off the Brazilian coast. Years later, when the time comes for them to lay their eggs, they then make their way back to Ascension Island, only 6 miles across and more than 1,400 miles away, with no land in between. The tracking of tagged turtles by satellite has shown that they can maintain a straight course over hundreds of miles and that they "have a surprising capacity to pinpoint specific targets during a long-distance journey in open seas, without movements indicating a random or systematic search." They continue on their bearing at night, even when the moon is not visible, and they compensate for drift due to currents.[12]

If sea turtles are captured and released far away from their normal range, they can still find their own way back. An early unplanned experiment, reported in 1865, involved a green turtle caught at Ascension Island and taken by ship as far as the English Channel, at which point the animal looked unhealthy and was thrown overboard. Two years later it was caught again at Ascension Island and recognized because it had been branded.[13] Turtles seem to have a magnetic sense,[14] but even the most sophisticated compass could not explain a navigational feat like this.

Most seasonal migrants move between breeding grounds and feeding grounds in a repetitive cycle, but some animals have no fixed route at all. Albatrosses, for example, wander over the oceans for vast distances with unpredictable routes in search of food, and yet they can still find their way back to their nesting places on midoceanic islands. Wandering albatrosses that nest on the Crozet Islands in the southern Indian Ocean have been tagged and tracked by satellite, and these studies have revealed that they can go foraging in any direction, and the outbound and inbound trips may be widely separated[15] (Figure 11.2). On their return trip, like the green turtles, they can approach their home island on a straight course as if they know exactly where it is, rather than searching for it. They cannot be locating their home by means of smell because they often return when there are crosswinds, or on routes that are upwind of the Crozet Islands.[16] Like the displaced turtles, they must be navigating toward their goal in a way that cannot be explained in terms of inherited programs and normal senses.

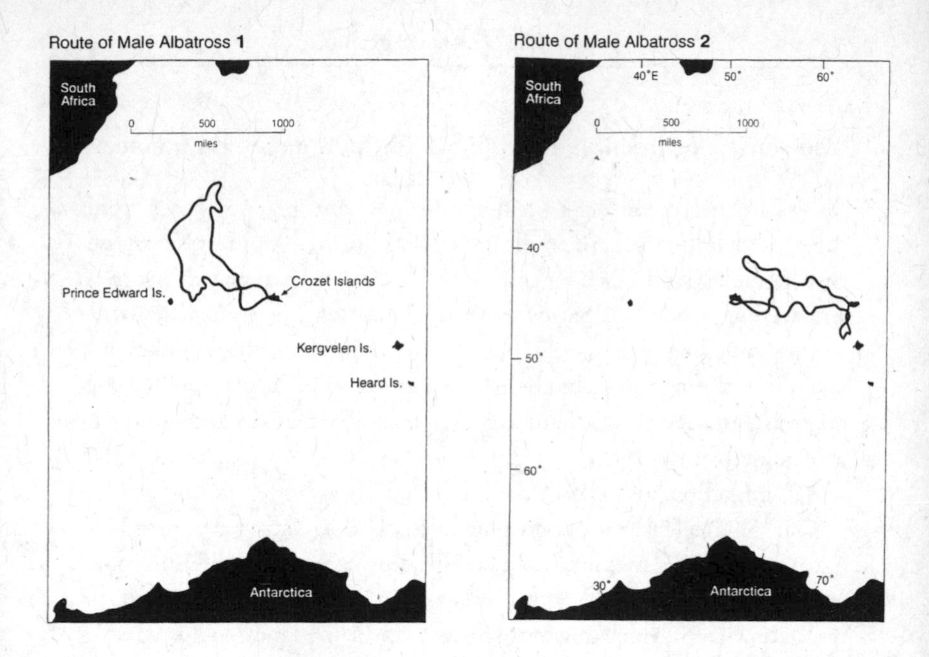


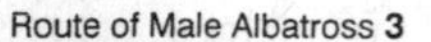

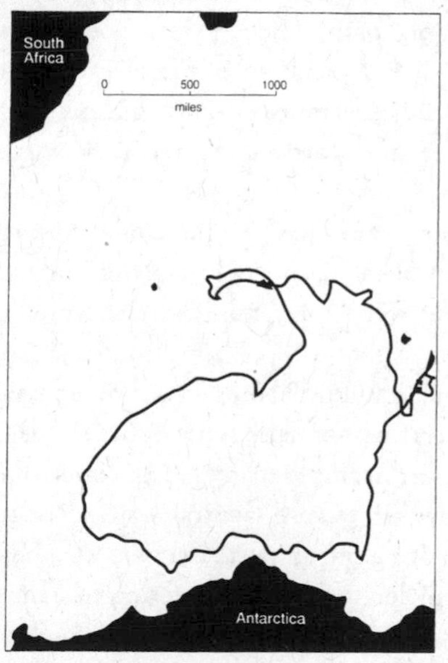


*Figure 11.2 Tracks of three wandering albatrosses in the southern Indian Ocean (after Jouventin and Weimerskirch, 1990).*

## The sense of direction, morphic fields, and ancestral memory

Just as a sense of direction in domestic pets and pigeons arises from their close links to familiar places, especially their homes, I propose that a similar kind of connection links turtles to their native beaches and feeding grounds, albatrosses to their nesting islands, and godwits to their breeding places and winter homes. These invisible connections take place through morphic fields—which act as attractors, pulling the migrating animal toward them as if by an invisible elastic band—and enable animals to navigate toward their goals. Such fields play an essential role in migration, just as they do in homing.

One of the features of morphic fields is that they have an inherent memory. This memory is transmitted by a process called morphic resonance, which causes a given organism, such as a migrating bird, to resonate with previous migrating birds of the same kind.[17] Thus when a young cuckoo sets off from England toward Africa, it draws upon a collective memory of its ancestors. This memory, inherent in the morphic field of its migratory path, guides it as it goes, giving it a memory of directions in which to fly, and an instinctive recognition of landmarks, feeding grounds, and resting places. This collective memory also enables it to recognize when it has arrived at its destination, the ancestral winter home.

Natural selection would have strongly favored birds that were sensitive to this ancestral migratory field and that migrated in accordance with it. Those that were not in tune with it would probably not have survived.

Migrations usually follow habitual routes, repeated over many generations. The migrating animals' sense of direction has a habitual sequence of stages. For example many North American migratory species are funneled in flyways toward Central America or the Gulf of Mexico and then diverge again into South America. On their return migration they are again funneled through Central America and the Gulf, and they follow one of several major routes northward—along the West Coast, for example, or up through the Mississippi basin.

Likewise some migratory species that breed in Western Europe, like swallows, are funneled toward the Strait of Gibraltar, the shortest sea crossing to Africa, and then fly across the Sahara. Populations of the same species breeding in Eastern Europe are funneled through the Bosphorus, where they cross the narrow strait between Europe and Asia.

In all these cases, the birds' sense of direction depends on the stage they are at in their journey. They do not set off in a straight line directed toward their winter or summer home, but rather follow flight paths toward traditional sea crossings, and often fly along coastlines and rivers.

Young birds that make the journey for the first time with no guidance from birds that have flown the journey before, like young cuckoos, have to rely entirely on this inherited sequence of directions and have no experience of the winter home or of the intermediate stages on the way.

After having spent some time in a wintering or breeding area, some migratory birds are able to navigate toward this place not only along the usual route but also from an unknown starting point. This establishment of a connection with the place is referred to as "site imprinting" in the scientific literature, but virtually nothing is known about how it actually works.[18] I suggest that this imprinting involves the establishment of connections with the place through a morphic field, which continues to connect the bird with that particular place even when it is far away.

## Experiments with migrating birds

In some classical experiments in the 1950s, conducted on a larger scale than any before or since, the Dutch biologist A. C. Perdeck investigated what migrating birds actually did when they were taken away from their traditional route. He displaced thousands of starlings and chaffinches by capturing them when they had actually started on their journey. These birds were ringed, taken hundreds of miles away, and released from a place they had never been to before. An international network of ornithologists sent in data on the recovery of ringed birds. The purpose of Perdeck's experiments was to find out whether experienced birds could navigate in a goal-directed way, as homing birds do, or whether they simply flew in a programmed direction. Perdeck explained the thinking behind his experiments as follows:

> The faculty of birds to orient themselves not merely in a particular compass direction but to a certain geographical position has been called "homing orientation," "complete navigation," or "true goal orientation." Its existence is doubtless proved by homing experiments in many species during the breeding period. Therefore it seems unlikely that this highly developed mechanism of orientation is not used during migration when it has so many advantages compared with one-direction orientation. . . . The homing experiments suggest that this faculty is developed especially in older birds, which have been already one or more seasons in the area of destination.[19]

In a series of experiments, repeated over several years, starlings migrating from the Baltic region to their usual wintering areas in England and northern France were captured at their autumn stopover sites in Holland. Eleven thousand captured birds were ringed and taken by plane to Switzerland, about 375 miles to the southeast, where they were released. Juvenile and mature birds were released separately. Normally starlings fly in flocks of mixed ages, the juveniles traveling with more experienced birds, but in this experiment, they were forced to make their own way.

The juvenile birds continued to fly southwest, the direction they would have taken from the place they were captured toward their wintering grounds in England. In other words, they followed a path parallel to the normal one. Some of them ended up in southern France and Spain. But the adults reoriented themselves (Figure 11.3) and found their way to the traditional wintering areas in England and northern France. In other words the adults were showing navigational behavior similar to that of homing pigeons, dependent on the connection they had made to their winter homes, their site imprinting.[20]

Perdeck obtained similar results with migrating chaffinches, also captured in Holland and released in Switzerland. The young chaffinches flew southwest, continuing in the same direction they would have flown if they had not been captured and displaced. But the adults,

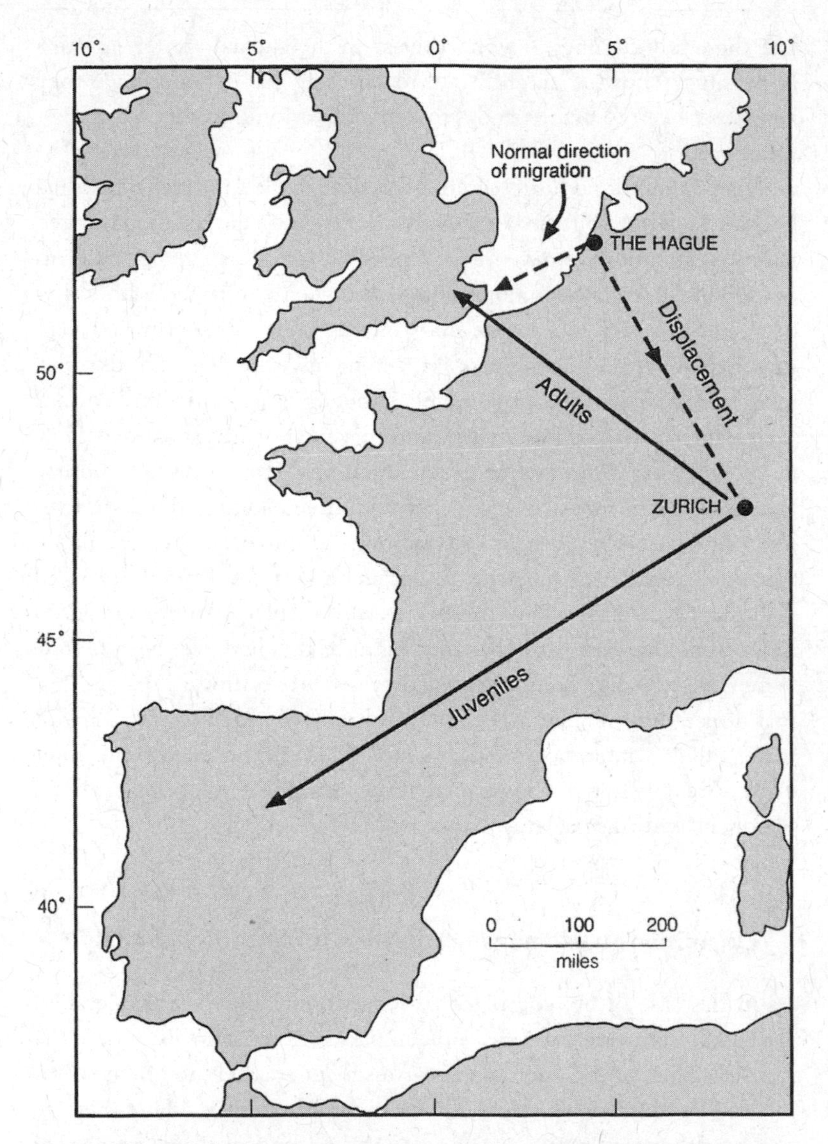


*Figure 11.3 The directions of migration of adult and juvenile starlings after their displacement from Holland to Switzerland. The adults flew to their usual wintering grounds in England, but the juveniles flew in the direction that would have taken them from Holland to England if they had not been displaced. Consequently they ended up in France or Spain (after Perdeck, 1958).*

like the adult starlings, flew northwest to their usual winter quarters in Britain,[21] showing an ability to reach their goal from a place they had never been to before, and flying in a direction different from their usual one.

When Perdeck did his research it was not possible to track individual birds as they flew, but since the early twenty-first century miniaturized radio-tracking devices have made it possible to follow even small birds as they move. Perdeck's findings have now been confirmed with a new degree of precision in a recent study on migrating white-crowned sparrows led by Martin Winkelski, of Princeton University. These birds breed in western Canada, and in September they migrate southward toward their wintering areas in southern California and northwest Mexico. They fly alone at night but congregate at traditional stopover sites to break their journeys. Winkelski's team captured fifteen adult and fifteen juvenile white-crowned sparrows at a traditional stopover site in Washington State and took them by air to the other side of the United States, to Princeton, New Jersey. Thus these birds were displaced more than 2,000 miles from anywhere that they or their ancestors had ever been before. When the juveniles were released they set off southward, in the normal direction of migration, but the adult birds headed west-southwest toward their wintering grounds. As in Perdeck's experiments, the adult birds were able to home toward areas they had lived before, even after a continent-wide displacement.

## Do animals have maps or are they pulled toward goals?

As in the case of homing pigeons, conventional theories of migration are based on the idea of maps and compasses. Some animals can use the sun's position, or the stars, or the earth's magnetic field to orient themselves. But compass directions alone cannot tell animals where they are or where their destination is. Humans need a map as well and a means of identifying their present position. But how could white-crowned sparrows have a map of the entire United States that enables them to work out that they had been displaced to Princeton and that they needed to fly west-southwest toward their wintering ground?

Some scientists have proposed that animals have large-scale magnetic maps that enable them to work out where on earth they are from the direction and dip of the magnetic field and the characteristic patterns of the field in different parts of the world, including local magnetic anomalies. For example, some have suggested that migrating turtles not only use a magnetic sense to detect their direction but also navigate by means of an inherited magnetic map that tells them where they are in the ocean, enabling them to change direction at appropriate places.[22] This would mean that the animals would have to inherit large-scale maps of the earth's magnetic field. Apart from the problem of imagining how such a map could be built up and inherited on the basis of random genetic mutations, there are two further difficulties.

First, the magnetic north pole wanders from year to year, the earth's magnetic map changes from decade to decade (Figure 11.1), and at irregular intervals the entire magnetic field of the earth reverses. How could a genetically programmed map cope with such changes? Proponents of this idea suggest that natural selection weeds out turtles that by chance have the wrong kind of map and favors those that by chance have mutant maps that fit the new magnetic field. But how would any of the animals survive a polar reversal? "Although responses to regional fields might be rendered useless during occasional periods of rapid field change associated with magnetic polarity reversals or excursions, these sporadic events do not preclude the evolution of magnetic responses during the intervening and usually much longer intervals when the earth's field changes more slowly."[23] This is a tenuous argument, since all the turtles would be wiped out if their maps are the wrong way around when a polar reversal occurs, unless by chance a few happened to have maps reversed just at the right moment.

Second, turtles taken far away from their normal range can still home, like the green turtle released in the English Channel that found its way back to Ascension Island in the South Atlantic. Green turtles normally live in tropical and subtropical waters. To explain the homing of turtles displaced far beyond their normal range, proponents of the magnetic map theory would have to suppose that they are born with an accurate magnetic map of large parts of the world where they and their ancestors never normally go.

What about the white-crowned sparrows taken from Washington State to Princeton, New Jersey? Winkelski's team could suggest only that they had some kind of inherited magnetic map, but he then admitted its inherent implausibility, saying the birds somehow work out where to go from the sky or by smelling their wintering area from more than 2,000 miles away: "Currently, magnetic clues seem the most likely candidates for the basis of a map stretching this far. However, the small difference in geomagnetic intensity across longitudes in North America makes magnetic intensity an unlikely candidate for distinguishing between the east and the west coast, and celestial or olfactory cues cannot be ruled out."[24]

This is the state of the art in homing and migration research. Over and over again, some researchers claim that they have explained animal navigation in terms of magnetism, or smell, or a celestial compass, and again and again these claims dissolve into contradictions and vague speculations about backup systems and the integration of cues from multiple sources. The truth is that after more than a century of research no one knows how animals find their homes or reach their migratory goals: This is one of the great unsolved problems of biology.

I have come to the conclusion that we will never understand long-distance animal navigation in terms of the known senses alone, even including a magnetic sense. Something very fundamental has been left out. It is as if a scientist tried to understand a mobile phone without knowing about radio waves. However much he analyzed the components of the phone or studied the earth's magnetic field in which the phone was operating, it would remain inexplicable.

What has been left out are the morphic fields that connect animals to their homes. Animals are pulled or attracted toward their goals through these fields, which underlie their sense of direction. They can use a magnetic sense, or a sun or star compass, or smell to help them keep on course, but their direction depends on a direct connection to their destination. This need not necessarily be their final goal but can be the next staging post on their journey. Many migrants follow traditional routes with stopovers where they rest or feed, and these too act as attractors on the way.

Because morphic fields contain an inherent memory, they also help to explain how migratory patterns are inherited and how they evolve.

## The evolution of new migratory patterns

Perdeck's experiment had a fascinating ending. In the spring, some of the young starlings that had been displaced and had found new wintering grounds in France and Spain returned to the countries of their birth around the Baltic. This showed that they were able to navigate to an area they already knew, although they had to reach it by a new route. And remarkably enough, the following winter some of these young birds returned to the new feeding grounds in France and Spain that they had adopted the previous year.[25] A new migratory cycle had been established in a single generation. Genetic mutations had nothing to do with it.

According to the morphic field hypothesis, new migratory pathways can evolve rapidly. Animals that have strayed or been blown off course might find new feeding grounds for the winter, as the young starlings did in Spain. When they homed to their native lands in the spring, they established a new migratory loop. And if this new pathway promotes the survival and reproduction of the animals that follow it, a new migratory race will come into being.

This kind of evolutionary process has actually been observed over the last fifty years with the European blackcap, a kind of warbler. These birds breed all over Europe. In the autumn blackcaps from Eastern Europe head toward the Bosphorus, in Turkey, and then make their way around the eastern Mediterranean to East Africa. Those from Western Europe traditionally migrate toward Spain, where some spend the winter, while others cross into Africa and winter in Morocco or West Africa. But since the 1960s a new migration route has developed: from central Europe to Britain, where many thousands of blackcaps currently spend the winter. About 10 percent of the breeding population in parts of Belgium and Germany now winters in Britain instead of Africa.

This new pattern of migration has become possible because of the progressively milder winters in Britain in recent decades, and also because so many British people feed birds during the winter, providing a new source of food unavailable in previous centuries. Also, this migration is much shorter and less hazardous than the usual journey to Spain or West Africa. Moreover, the birds that winter in Britain tend to return to the breeding grounds sooner than those that have to travel farther, enabling them to pair with each other earlier, occupy the best territories, and produce more offspring.[26] Thus natural selection favors this new migratory habit, and a new race of blackcaps is emerging.[27]

According to the conventional genetic programming theory, the evolution of a new migratory pathway would depend on chance mutations that affect the genetic programming. Then these mutant genes would have to be favored by natural selection over many generations for a new race to emerge.

By contrast, if migratory routes are more like habits that depend on an inherited memory, new races can emerge rapidly, as the example of the blackcaps shows. To start with, this process need not require any genetic mutations at all. Blackcaps might originally have come to Britain because they were blown off course from their normal migratory route toward Spain rather than because of mutations in hypothetical migration-programming genes. Likewise, the new migratory pathway of starlings from the Baltic to Spain rather than Britain happened because a Dutch scientist kidnapped them and took them to Switzerland in an airplane, not because of mutant genes.

Under natural conditions, new patterns of migration could arise whenever animals are carried away from their habitual route and are lucky enough to find themselves in a new feeding ground. If favored by natural selection, this pattern could be repeated over the generations, as in the case of the blackcaps. The new destinations and the migratory pathways themselves would not be coded in the genes but would be remembered through the morphic fields.[28]

This conclusion is supported by recent genetic research on green turtles. Although there are some genetic differences between races that breed far away from one another and follow widely different migratory routes, these differences are so small that researchers on this subject have concluded that "migratory routes to specific destinations such as Ascension Island may not be genetically fixed or 'instinctual.' . . . Early learning, rather than 'hard-wired' genetic behavior, would allow a more flexible response to altered nesting conditions, so that new migrational pathways could be established in a single generation."[29]

*Chapter 12*

# Animals That Know When They Are Nearing Home

Many pet owners who travel with their animals have surely noticed a puzzling kind of behavior shown by many dogs, cats, horses, and other species. Their animals often seem to know when they are nearing their destination, even when they cannot see out of the vehicle in which they are traveling.

As we have seen, homing and migration depend on morphic fields that pull or attract the animals toward their destination and underlie their sense of direction. In this chapter I suggest that this morphic field hypothesis can also shed light on the way pets know when they are nearing their destination. In some cases, this behavior depends on the morphic fields that connect animals to particular places. In other cases, it seems to depend on the people the animal is traveling with, and may involve telepathy.

## Animals traveling in cars

Our cat, Remedy, unlike most cats, enjoyed traveling by car. For most of the journey she would sleep on a rug in her carrying basket. We left the door of the carrier open so that she could come out whenever she liked.

A mile or two before we arrived home, whether we were traveling by day or night, Remedy would wake up, come out of her basket, pace around the car, and show obvious signs of excitement. My wife was the first to notice this. I am sorry to say that I was at first dismissive of this phenomenon, arguing that it must just be a matter of chance. I was afflicted by a closed-minded skepticism that led me to ignore or deny behavior that seemed to have no immediate explanation. But in the end, the evidence became overwhelming. The cat did seem to know when we were nearing home. But how?

Was this because she smelled the home area? Possible, but unlikely, because this seemed to happen just as often in cold weather when the windows were closed as in hot weather when they were open. Did she recognize familiar bumps, twists, and turns in the road? Possibly, but how could she be so familiar with the details of different routes through London, with variable patterns of movement, depending on traffic lights and road congestion? Was she responding to some other quality of the home environment? Perhaps, but what could it be? Did she somehow pick up our own anticipation, even when we were not aware of any change in our behavior? If so, how?

I still do not know the answers to all of these questions, but I have found that many other people have noticed similar behavior in their animals. We have more than sixty accounts of it on our database, and taken together they enable us to narrow down the possibilities.

No doubt when wild animals are making their way home, they recognize familiar landmarks, smell familiar odors, and hear familiar sounds. Without this information from the environment they would be lost. But a wild animal is in a very different situation from a dog or cat asleep in a car, or a horse in a horse trailer. These animals are being transported whether they like it or not and with no choice as to where they are going. This must be a very unusual situation for animals to be in. Wild animals are rarely carried around, although sometimes victims are carried off alive by predators, and young animals are sometimes moved by their elders, as when mother cats move their kittens by the scruff of the neck. But I can think of no parallel in the natural world for the transportation of domestic animals in vehicles, except perhaps for fish being carried along by currents, or birds by winds.

## Arriving at familiar destinations

Cats most commonly react when they are nearing home, just as Remedy did. Many dogs show similar signs of anticipation, and in most cases it seems unlikely that they know where they are by seeing landmarks. On car journeys they usually lie down, below the level of the windows, and go to sleep. Looking out of the windows is a result of their waking in anticipation, not a cause of it. And many dogs and cats react even when they are traveling at night.

Likewise, some horses when transported in vehicles seem to know when they are nearing home, and several miles beforehand start fidgeting, whinnying, pawing, stamping, or showing other signs of restlessness.

It is rarer for bonobos to travel by road, but those at Twycross Zoo in Warwickshire used to do so when they made a series of TV commercials in which they were dressed up as humans and acted out a sketch. Molly Badham, their trainer, found that on the return journey they always seemed to know when they were nearing the zoo. "Probably a mile away, they would wake up and know they were coming home. How they knew, I don't know—it was pitch black and they couldn't see out anyway—but they would wake up and start getting excited."

Many animals show similar reactions when they visit familiar destinations other than their home. Alice Palmer of Chicago, Illinois, told us about Tasha, her poodle: "When we went to visit my son, one hundred twenty miles away, and were within five to eight miles of his house, Tasha would jump up from her sleeping position in the back seat, sniff at the window in the back seat, and excitedly watch until we reached the house." Perhaps Tasha was responding to familiar smells along the road. But what happens when animals are taken by unusual routes?

## Traveling by unusual routes

Several people have deliberately tried taking a different route to find out how the animal reacts. Jenny Mardell of Bath, England, found that her dog, Mandy, always got excited when they were nearing her parents' home in London: "We could never work out how, and would try different routes to the house. She always knew and would get hysterical in the car." Likewise, Geneviève Vergnes found that when she visited her parents in Paris, "the dog would wake up at about seven kilometers from their house and scratch the dashboard while 'singing.' We deduced she knew the way and tried different routes—the peripheral quais, the Champs Élysées, or by way of the suburbs—the duration of the route being different. She slept, and always at about the same distance she would scratch the dashboard and sing!"

Sometimes people are strongly motivated to try to prevent their dog from waking up and showing its excitement. This was true of a London couple when their twin children were babies. When traveling home by car, their Labrador Retriever and the babies would go to sleep. As they neared home the dog would wake up and move around excitedly, waking the twins, who would start to cry. The parents tried driving home by a variety of devious routes, but they failed to deceive the dog. He got excited and woke the babies anyway.

Experiences like these show that at least some animals know when they are nearing their destination no matter what route they are taking. Does this depend on some quality of the place that they can detect, even when asleep, when they are still miles away? Or does it depend on some influence from the people in the car?

After studying dozens of cases, I have concluded that in some animals the place itself plays the more important role, but in others the pets are picking up the expectation of the people in the car. And probably both kinds of influence can work together.

## Familiar and unfamiliar places

In most cases, the reaction of animals occurs only before arriving home or at some other familiar place. They do not react before reaching unfamiliar places. This suggests that they detect something about the places themselves and that their reactions depend on memory. This is particularly clear from the observations of Joséa Raymer, of Aldermaston, Berkshire, the owner-driver of a truck. She always takes at least one German Shepherd with her for protection:

> I can drive for up to four and a half hours on all sorts of road conditions (stopping for traffic lights, etc., crawling through slow traffic, or on fast motorways) and the dogs will stay asleep, but as I near my delivery point or overnight stop the dogs will react in direct relation to whether they have been there before, how many times, and whether they were let out for a run. If I have never delivered there before, or if it was a place not suitable for letting them out, they will take no notice until I have stopped, reversed into the loading bay, or opened the door. If it is a place we visit often, or especially a night stop where they can expect some fun, they will be up and getting excited while [we are] still on the public road, where you would not expect the motion of the truck to be any different from any other part of the road.

In cases such as this, memory of the place seems far more important than the behavior or thoughts of the person. But what is it about the place that they respond to? We have already ruled out landmarks, since animals that are asleep, or at least lying down, cannot see them out of the window, and they still react when traveling in darkness.

So could it be smell? This is the most obvious possibility, and in some cases it may be the best explanation. But in most cases it does not fit the facts very well. The smell theory would predict that animals should react sooner in warm weather than in cold, since the vaporization of substances is greater at higher temperatures. They should also react sooner and more strongly when the windows are open than when they are closed. And their reactions should depend on the wind direction, occurring much farther away when the destination is upwind than when it is downwind. There is no hint in any of the accounts that I have read or heard that this is the case, nor did I observe any such influence of wind direction, temperature, or open windows on the reactions of our own cat.

These reactions may be related to the sense of direction discussed earlier. This sense of direction is of primary importance when animals actively make their own way. It has evolved in the context of their finding the way home or traveling to other familiar locations, and it also plays an essential role in migration. I suggest this sense of direction depends on morphic fields, through which animals are attached to familiar places. These fields enable them to find these places, navigating over unfamiliar territory, and may also enable them to recognize when they are getting near a familiar place when someone else is doing the navigating.

## Reacting to people rather than places

When nearing their destination, the people in the car may feel relief or anticipation, and animals may sense these changes and wake up. If they are about to arrive at a familiar place, it is hard to separate the effects of the people from the effects of the place itself. But when they are traveling to a place with which the animal is unfamiliar, the animal can have no memory of the place, and the only possible clues could come from the people in the car.

In some cases, animals do indeed seem to react before arriving at a place they have never been to before. For example Jenny Vieyra of Leighton Buzzard, Bedfordshire, has a cat that has always known when they are nearing home, "standing up in his basket, meowing frantically, trying to find a way to get out." He also knows when they are about to arrive at the houses of friends or members of the family, or at boarding kennels where he has stayed before. All this could be a matter of memory. But Jenny recently moved to a new house 50 miles away, and when she took the cat there for the first time in his life, he reacted with his usual excitement before they arrived. He must have been picking up the anticipation of his owner.

Some people who have seen this kind of reaction repeatedly are convinced that their animals are telepathic. Michaela Dickinson-Butler of Burton-on-Humber, for instance, thinks her Border Terrier can read her thoughts. "He knows exactly when we will stop the car and starts barking and whining before arrival at the destination, even if he's never been before, and he doesn't look out of the window, and we are silent."

Peter Edwards of Essex breeds Irish Setters and often attends dog shows. When he is going home with his dogs in the car, they usually wake up and get excited some fifteen minutes or more before arriving. They do the same before arriving at a showground. Sometimes he takes the dogs to shows in new places, and they still react fifteen to twenty minutes in advance. Smelling other dogs already at the showground seems an unlikely explanation, because the response occurs many miles away, and it does not seem to depend on the wind direction. Moreover, Peter Edwards has found that they react even when he arrives at a showground early, when very few other dogs have arrived. "I think they are picking it up from me," he says. He has inquired about other owners' experience of taking their dogs to shows and has found that the dogs' anticipation is quite common. "It's almost as if they can read minds," he says.

This phenomenon has also been studied by Elizabeth Marshall Thomas, who describes how her dingo, Viva, knew when they were about to reach their destination even if she had never been there before. Thomas tried to find out how she did it. The bumpiness of the road when the car went off highways onto country byways was a clue, but not the only clue, because the destination was often reached over many miles of unimproved surface, and Viva still knew. Likewise the repeated turns onto smaller roads and driveways before the end of a trip were a clue, "but often Viva brightened for arrival before the car began turning." When Thomas realized that Viva was accurately predicting most arrivals, she tried to make sure she herself was not giving a clue by speaking or doing something different. "I believe I succeeded in hiding my feelings. [Viva] still knew, though, and by the end of her life I was no closer to figuring out how she did it than I had been at the beginning. Dog ESP? Perhaps."[1] The ability of an animal to pick up a person's anticipation of the end of a journey is not very different from picking up other kinds of thoughts and intentions. Whether this depends on unconscious behavioral cues or on telepathy is another question. To find out, it would be necessary to do special experiments.

## A simple experiment to test for telepathy

Animals that seem to be reacting to clues from people in the car may actually be reacting to subtle behavioral changes, body language, spoken clues, or other signs detectable by the normal senses. One way of testing this possibility is for the animals to be carried in the back of a van, while their owner rides in the front. The animals could either be observed by a person riding with them in the back of the van, who does not know the destination of the journey, or their behavior could be recorded automatically by a video camera mounted in the back of the van.

In some preliminary experiments, conducted with the kind cooperation of the Manchester (England) Police Dog Unit, we have established that it is quite feasible to videotape the behavior of dogs in vans. But police dogs are accustomed to traveling in vans, and animals not familiar with this form of transport may take some time to get used to it.

With the animal in the back of the van, and the recording system in operation, the animal's owner then drives to destinations with which the animal is unfamiliar. Does the animal still show signs of anticipation? If so, the reaction cannot be explained in terms of a memory of the place, through seeing its owner's body language, or through other sensory clues. By a process of elimination, telepathic communication would be the most probable explanation.

*Chapter 13*

# Pets Finding Their People Far Away

In 1582, Leonhard Zollikofer left his native St. Gall, Switzerland, to go to Paris as ambassador to the court of the French King Henri III. He left behind his faithful dog, aptly named Fidelis. Two weeks later the dog disappeared from St. Gall. Three weeks after that he rejoined his master at the court in Paris, exactly at the time when the Swiss ambassadors were being led in to an audience with the king. The dog had never been to Paris before.[1] How did he find his master so far away from home?

It would be easy to dismiss this as a fanciful tale but for the fact that there are many such stories and even more heroic examples of canine devotion. One of these dates back to the First World War: Prince, an Irish Terrier, was devoted to his master, Private James Brown, of the North Staffordshire Regiment, and was inconsolable when this young man was posted to France in September 1914. Then one day he disappeared from his home in Hammersmith, London, and to everyone's amazement turned up at Armentières a few weeks later and tracked down his master in the trenches in a frenzy of delight. Because no one could believe the story, the commanding officer had man and dog paraded in front of him the next morning. Evidently Prince had attached himself to some troops who were crossing the English Channel, and had then found his way to his owner. He became the hero of the regiment and fought beside his owner for the rest of the war.[2]

In both these cases the dogs were not homing or going to another familiar place. Such behavior, if it really happens, cannot be explained in terms of a sense of direction—at least not a sense of direction that depends on any quality of the destination itself. Rather, the animals somehow knew where to find the people to whom they were so attached.

We have seen how bonds between animals and people can enable an animal to pick up intentions and calls at a distance. Some animals seem to know when their owners have died, even if they are far apart, or when they have had an accident. These various phenomena can be described as telepathic.

But could a telepathic link be directional? Can these bonds connect the person and the animal in a directional way, as if they are joined by an invisible cord?

We have in fact already encountered evidence for directional information in the bonds between animals and people. Some cat owners have found themselves drawn toward their lost cat in a way they cannot explain. They somehow know in which direction to look. Also, some animals know not only when their owners are coming home but also from which direction they are coming. They may wait at one or another side of the house, depending on the direction from which a person is approaching.

In this chapter I look at cases of animals finding their people in unfamiliar places. If some animals do this in a way that cannot be explained by chance, sight, hearing, or smell, then there must be two different kinds of directional sense: a sense for places and a sense for people or animals. My own hypothesis is that both kinds of bonds depend on morphic fields. I suggest that the morphic fields linking animals to places and also to people are indeed directional. But whereas a sense of direction for places is a familiar idea, a sense of direction for people or other animals is much less familiar. Finding people or animals at a distance is much rarer than finding places. Nevertheless, there are 119 such cases on our database, most concerning dogs.

I begin by looking at these spontaneous cases to see if there are any common patterns and to examine how convincing the evidence seems to be. There is generally no possibility of doing experiments on this phenomenon because pet owners are rightly reluctant to risk losing their animals. Practically the only source of evidence is from unplanned real-life events.

Chance and smell are the most important alternative explanations. Chance accounts for those cases in which an animal simply searched at random and happened to be successful. Animals can, of course, also track a person using their sense of smell. These are important arguments, for if chance or smell can explain the evidence, we have no need to postulate the existence of a mysterious invisible bond that could somehow pull an animal toward a person.

## Could the animals have found their people by smell?

Some of the cases on our database involve dogs or cats finding people in new homes less than two miles away from their previous home. Because the journeys of removal were made by car or van, tracking the person by following an odor trail along the road does not seem very likely. Nevertheless, smell and chance remain possible explanations, especially as the people may have left odor trails in the vicinity of the new home that an animal exploring at random could have picked up.

Sometimes dogs find their owners at their place of work, several miles away, never having been taken there. When the owners travel to work by car, tracking by scent seems a very unlikely explanation. Even if a dog or cat could follow the scent trail of a car's wheels, distinguishing these wheel tracks from those of many other vehicles, to put this ability into practice would involve the hazardous procedure of sniffing along roads. Even if traffic was sparse, in order to avoid these animals drivers would have to notice them, and might well remember seeing them. I know of no reports from motorists of such road-sniffing animals.

Patricia Burke lived on a farm on the Isle of Skye, Scotland, and went to work 6 miles away in Portree, leaving her terrier at the farm. She did not drive directly to work but first went 3 miles in a different direction to pick up a colleague. One morning, to her surprise, she found the dog sitting outside her workplace in Portree. "How did he know I worked there?" she asked. "He'd never been to Portree."

A skeptic could argue that the dog detected the smell of the workplace, having become familiar with it from smells on the owner's clothing. But this would not explain how the dog found its way to Portree to start with, unless the sense of smell is imagined to extend far beyond the range for which there is any evidence.

In any case, the smell theory cannot apply to dogs that find their owners in places that the owners themselves have never been to before. The dogs could never have smelled the odor of that place on their owners' clothing or hair. Nevertheless, there are several cases of dogs finding their people in a house they are visiting for the first time, in a hospital to which they have been rushed, or in an unfamiliar pub.

When Victor Shackleton was a teenager he kept a Greyhound, Jonny, to whom he became very attached. But the dog was unwelcome in his family home owing to lack of space. Much to his dismay, his father arranged to sell Jonny to a nephew who lived in a mining community in Yorkshire. Father and son traveled there by train from their home in Cheshire and handed the dog over to his new owner. Having tied the dog up in his backyard, he insisted on a farewell drink before they began the long journey home. They squeezed into a battered van and drove to a pub frequented by Greyhound enthusiasts. Victor explains what happened next:

> I sat mournfully as they talked. I thought of Jonny locked in that strange backyard, lonely, abandoned. The pub filled up and then suddenly the door was flung wide open, I was smothered with body and paws. It was Jonny, a piece of broken rope dangling from his collar. Everyone who saw Jonny arrive at the pub could not believe it. When my dad and his nephew explained the sequence of events that evening, it hit us all—how the hell did the dog find us at this pub three miles away and in a place he had never been before? At once, expert dog handlers in the pub insisted that Greyhounds cannot hunt by smell; they rely on sight entirely.

There seems no possible explanation for Jonny's behavior in terms of sight or hearing, and the smell theory is very improbable, as the experts pointed out.

Sometimes the ability of dogs to find their owners has actually saved the person's life, as in the case of Uri Geller, the celebrated spoon bender.

Around the age of fourteen, Geller was living in Cyprus and loved to explore the hillside caves above his school near Nicosia. Usually he went with friends and stuck to tested paths. This time he did neither.

> I got lost. Deep in the caves, cold, wet, and terrified, I spent two hours hunting with a failing flashlight for a way out. Finally I curled up into a ball and prayed to God that someone would find me before I starved to death, as two of my schoolmates had. I'll never know how my dog, Joker, reached me. I'd left him miles away at my stepfather's hotel. But huddled in the darkness I heard him barking—and suddenly his paws were on my chest and he was licking my face. Joker knew the way out, of course. It was as if my prayers had summoned him.[3]

Even if the dog was able to locate Geller in the cave by following a scent trail, this would not explain why he went to the cave in the first place and did so at exactly the time he was needed.

In most cases we have no information about the way a dog behaves on its journey. What route did it follow? Was it sniffing as if following a scent trail? Fortunately, I have received one account in which two dogs were accompanied all the way on their journey. Dr. Alfred Koref and his wife kept two Dachshunds at their home in Vienna, Austria. When they went out for the evening they left them at the house of their maid. In the morning, Dr. Koref picked them up from the maid's house and set off to walk them home while his wife drove on to visit some friends who lived 2 miles away. "Instead of going home, though, they pulled me through little streets that were unknown to them, until we came to the big street that my wife had driven along. Then they raced along this street and arrived at the block of buildings where our friends lived. The dogs had never visited them. Still, they went into the right entrance, climbed upstairs straight to the door where my wife could be found. She was quite surprised at our appearance."

Although in this case the finding of the right entrance and right apartment could perhaps be explained by smell, the journey through unknown streets and the racing along the pavement to the right buildings cannot be accounted for in this way. The smell theory becomes even more implausible when animals find their people over tens or hundreds of miles.

## Finding people over great distances

The greater the distance over which the animals find their person, the less plausible the random search theory and the smell theory become. Out of 120 cases on my database, only five involve animals finding people at distances more than 50 miles. Luckily, this subject has already been researched by one of the pioneers of parapsychology, Joseph B. Rhine, of Duke University, North Carolina. In the 1950s, he identified this phenomenon and named it psi-trailing, "psi" referring to the psychic nature of this ability.[4] Rhine and his colleagues built up a collection of cases through appeals in newspapers and magazines and from reports in local newspapers, through which some of the most remarkable cases first came to light. Wherever possible, they followed up these accounts with interviews and visits to obtain more details.

In 1962, Rhine and Sara Feather published a summary of their investigations. From their initial collection of data, they first screened out cases for which too little detail was given or where the people involved could not be identified and located. That left fifty-four cases of what appeared to be psi-trailing in animals, twenty-eight with dogs, twenty-two with cats, and four with birds.[5]

They then excluded all those cases in which the animals that found their people could not be conclusively identified as the former pet, as opposed to a stray resembling it that had found them by chance. Some were excluded because no further inquiries were possible, either because the events happened too long ago, or because the people were unable or unwilling to cooperate. And all those involving distances under 30 miles were ruled out, to reduce to a very low level the probability of the animal making a merely random search. After all these exclusions there remained a number of quite impressive cases that Rhine and Feather described in detail.

In one of these, Tony, a mixed-breed dog belonging to the Doolen family of Aurora, Illinois, was left behind when the family moved more than 200 miles to East Lansing, around the southern tip of Lake Michigan. Six weeks later Tony appeared in East Lansing and excitedly approached Mr. Doolen on the street. The rest of the family recognized

Tony too, and he them. His identity was confirmed by the collar, on which Mr. Doolen had cut a notch when they were in Aurora.

The most remarkable cat story centers on Sugar, a cream-colored Persian belonging to a California family. When they were leaving California for a new home in Oklahoma, Sugar jumped out of the car, stayed for a few days with neighbors, and then disappeared. A year later the cat turned up at the family's new home in Oklahoma, having traveled more than 1,000 miles through unfamiliar territory. Sugar was recognizable not only by her appearance and familiar behavior but also by a bone deformity on her left hip, which Rhine himself examined.

Pigeon number 167 was identified by the number on its leg ring. The owner of the pigeon was a twelve-year-old eighth grader from Summersville, West Virginia, where his father was sheriff. This racing pigeon had stopped in his backyard; the boy had fed it, and it had stayed and become his pet. Some time later, the boy was taken for an operation to the Myers Memorial Hospital at Philippi, 105 miles away by road (70 by air) and the pigeon was left behind at Summersville. "One dark, snowy night about a week later," according to Rhine and Feather, "the boy heard a fluttering at the window of his hospital room. Calling the nurse, he asked her to raise the window because there was a pigeon outside, and just to humor the lad, she did so. The pigeon came in. The boy recognized his pet bird and asked her to look for the number 167 on its leg, and when she did so she found the number as stated."

In addition to this collection of cases by Rhine and Feather, similar stories have been reported from many different countries. In France, for example, a two-year-old Sheepdog was left in Bethune, in northeastern France, while his owner set off on a career as a traveling construction worker. One day when he was working in Avignon, 500 miles away, he was told of a stray dog behaving strangely in the vicinity. He went to investigate and was almost bowled over by his dog, overjoyed at being reunited with his master.[6]

There is even a story about a pet magpie, reported by Mrs. M. Johnson, a teacher in Lund, Sweden. One day a magpie flew in through the open window of a corridor in her school and perched on the shoulder of one boy in a group of about forty children. He exclaimed, "It's our summer bird!" Then he explained that his family had spent the summer in a cottage about 50 miles away, where they had kept the magpie as a pet. When they moved back to the city, they left the bird behind. It was so clear that the bird knew the boy that his teacher excused him from school so that he could take it home.[7]

The many stories of this kind have convinced me that animals can indeed sometimes find their people in a way that cannot plausibly be ascribed either to smell or to chance.

## Dogs finding their owners' graves

Stories about pets finding their owners are not confined to living people. Over and over again, we hear tales of dogs finding their owners' graves. I find these stories baffling. At first I assumed that the bond between pet and owner would be dissolved when the person died. I took it for granted that it would not connect the animal to the person's dead body. But this assumption now seems ill-founded, given the many stories of grave-finding dogs. Consider this one from Joseph Duller of Graz, Austria:

> My father-in-law had a small farm and on it he kept a watchdog, Sultan. One day my father-in-law became ill and was taken to the hospital by ambulance. A few days later he died and then he was buried in the local graveyard, five kilometers from the farm. Several weeks after the burial the dog was not seen for days. This seemed strange to us, as Sultan never used to stray. But we did not make much of it, until one Sunday a former employee came along, who lived near the graveyard. She told us: "Imagine, when I went across the graveyard the other day, Sultan lay at your family grave." I cannot fathom how he could have found the way all these five kilometers. There were no footprints of his former master that he could follow. And he had never been taken to the graveyard, not even to the fields, since he had to keep watch at the house. How is it possible that he found his master's grave?

Presumably members of the family had been to visit the grave, and could perhaps have left scent trails leading there. But they would presumably have left scent trails to many other places as well. So if Sultan followed them, why did he go specifically to the cemetery, and how could he have known about bodies being buried in graves?

Why should mourning dogs be attracted to their owners' graves at all? The only possible reason seems to be that their attachment persists after their person's death and continues to be focused on the owner's body. We have already seen several striking examples of faithful dogs that stayed by their dead owners' bodies or kept vigil by their graves. The bonds that attach animals to their owners are not necessarily dissolved by the owner's death. I would expect this connection to be much weakened or dissolved by cremation, when nothing remains of the body except ashes. And all the cases I know of concern burials; I have never heard of animals being drawn to crematoriums or to places where ashes are scattered.

This attachment of dogs to the dead bodies of their much-loved people may seem strange. But many people, after all, retain a remarkably strong connection to the bodies of their loved ones. These continuing connections have given rise to tombs and gravestones, and simple acts of devotion like the placing of flowers on graves. After all, cemeteries are visited not just by mourning dogs but by human beings, too. And the burial places of particularly important people, such as saints and national heroes, become places of pilgrimage for thousands, even millions. If we understood better why we visit graves ourselves, we might gain more insight into the bonds that continue to link some dogs to the dead bodies of their owners.

## Animals finding other animals

Occasionally reports appear in newspapers about farm animals that are separated from their young and that later succeed in finding them again. Here is one example:

> Blackie, a two-year-old heifer broke away from the farm she had been sold to and, walking seven miles through strange country, homed in on the new farm that her calf had been taken to. The story started when the heifer and her calf were sold separately in Hatherleigh

market in Devon. The mother was sent to Bob Woolacott's farm near Okehampton where she was bedded down for the night with a supply of hay and water. But her maternal instinct led her to break out of the farmyard, and over a hedge into a country lane. Next morning she was found seven miles away reunited and suckling her calf at Arthur Sleeman's farm at Sampford Courtenay. Mr. Sleeman was able to identify Blackie as the mother by the auction-labels still stuck on their rumps.[8]

We have checked the details of this story by interviewing the people involved. Mrs. Mavis Sleeman told us that her husband put the newly purchased unweaned calf with the other calves. "The next morning my sister-in-law saw a cow coming down the lane. It came straight round to our building, where the calf was, this was at eight in the morning. It obviously wanted to get in there so she opened the door and let it in and the cow walked immediately over to her calf to suckle her."

Here is a similar report from Russia: "Caucasian farmer Magomed Ramazhanov was a little surprised when one of his cows went in search of her calf, sold earlier to a farmer in a neighboring district. Originally fearing that the creature had been killed by wild predators, Magomed eventually found his mild-mannered milker reunited with her offspring—30 miles from home."[9]

As far as I know there has been practically no research on the way animals find each other at a distance. One of the few investigators who has paid attention to this question is the American naturalist William Long. In his pioneering studies on the behavior of wolves in Canada, he paid particular attention to the way the members of the pack were linked together even when they were far apart. He found that wolves that were separated from the pack seemed to know where the others were. "In the wintertime, when timber wolves commonly run in small packs, a solitary or separated wolf always seems to know where his mates are hunting or idly roving or resting in their day-bed. The pack is made up of his family relatives, younger or older, all mothered by the same she-wolf; and by some bond or attraction or silent communication he can go straight to them at any hour of the day or night, though he may not have seen them for a week, and they have wandered over countless miles of wilderness in the interim."[10]

Through long periods of observation and tracking, Long concluded that this behavior could not be explained by following habitual paths, tracking scent trails, or hearing howling or other sounds. For example, he once found a wounded wolf separated from the pack; it lay in a sheltered den for several days while the others ranged widely. Long picked up the tracks of the pack in the snow, followed them while they were hunting, and was near them when they killed a deer. They fed in silence, as wolves commonly do, and there was no howling. The wounded wolf was then far away, with miles of densely wooded hills and valleys between him and the pack.

> When I returned to the deer, to read how the wolves had surprised and killed their game, I noticed the fresh trail of a solitary wolf coming in at right angles to the trail of the hunting pack. . . . I picked up his incoming trail and ran it clear back to the den from which he had come as straight as if he knew where he was heading. His trail was from eastward; what little air that was stirring was from the south, so that it was impossible for his nose to guide him to the meat even if he had been within smelling distance, as he certainly was not. The record in the snow was as plain as any other print, and from it one might reasonably conclude that either the wolves can send forth a silent food call or else a solitary wolf might be so in touch with his pack mates that he knows not only where they are but also, in a general way, what they are doing.[11]

These connections may be a normal feature of animal societies, even though we have hardly begun to understand how they work. The bonds between the members of a social group such as a wolf pack may not only enable them to know of the others' activities and intentions at a distance, but also provide them with directional information. And if wolves and other wild species have such abilities, then the ability of pets to find their owners and of owners to find their pets can be seen in a much wider biological context.

## Bonds to members of the social group and bonds to places

If some animals have the ability to find other animals or their human companions, it seems unlikely that nonhuman animals alone would be possessed of this power. We might expect to find the same kinds of phenomena in people, though probably to a lesser degree.

I would expect, therefore, that there might be stories about people who have found other people in a remarkable way, without knowing how. I would expect that these stories would principally be about people who were strongly attached to one another, such as parents and children or husbands and wives. I would also expect that in traditional hunting and gathering cultures, where finding both places and people was necessary for survival, these abilities may have been cultivated and encouraged, and may have been far better developed than they are in modern industrial societies.

Just as a sense of direction must depend on a connection between the animal and a place, so finding a person who has moved away must depend on a connection between the animal and a person. And just as the animal-place connection can be compared to a magnetic attraction or to a stretched elastic band, so can the animal-person connection. Both animal-person and animal-place fields, like magnetic fields, contain directional information.

The magnetic field of the earth contains directional information, which is why you can use a compass to find out where north is. In the technical language of science, magnetic attractions and repulsions are *vector* phenomena, which have a direction as well as a magnitude. By contrast, a *scalar* quantity—for example temperature—has magnitude but not direction.

These animal-person connections and animal-place connections are vectors, with both direction and magnitude. Only if the pull is strong enough does the animal set off on its journey and keep going in spite of distraction and adversity. And only if the pull has a direction does the animal know which way to go.

The idea of morphic fields linking animals to other members of their social group provides a basis for understanding both telepathic communication and a directional pull toward animals or people. The idea of morphic fields linking animals to particular places provides a basis for understanding the sense of direction as expressed in homing and migration. Thus the morphic field hypothesis may account for a wide range of unexplained powers of animals, both telepathic and directional.

But this hypothesis may not prove so helpful in one major category of unexplained perceptiveness—premonition. This is the subject of the following two chapters.

Part VI

# Animal Premonitions

*Chapter 14*

# Premonitions of Fits, Comas, and Sudden Deaths

A premonition is a warning in advance, a forewarning. Some premonitions seem to depend on telepathy, as when animals know in advance when their people are coming home. In some instances the animal may have detected odors, sounds, electrical changes, or other physical stimuli. But others may involve precognition, which literally means "knowing in advance," or presentiment, "feeling in advance."

Precognition and presentiment are more mysterious than other kinds of premonitions because they imply that influences can travel backward through time from the future to the present and from the present to the past. Such a notion defies our certainty that cause always precedes effect. It also flies in the face of our idea of the present, suggesting there are no sharp divisions between future, present, and past.

Can we avoid these problems and paradoxes? Can premonitions be explained without precognition or presentiment? Some kinds can; perhaps others cannot.

In this chapter I discuss the ability of animals to warn us of internal dangers, such as impending epileptic fits. But first it is important to consider the biological context of warnings and alarms.

## Danger, fear, and alarm

Fear is related to danger. The word itself originally meant danger.

We all know fear from our own experience. We feel that something bad is imminent. We become more alert. Our hearts beat faster. A surge of adrenaline prepares us to react. Our faces become pale; our hair stands on end; in extreme cases we may tremble with terror, and our sphincters may relax. Fear is an emotion we share with nonhuman animals and that we can easily recognize in them. And it is of obvious survival value, especially in relation to predators.

Fear triggers defensive behavior in any animal capable of defense. It can set animals running, diving, hiding, freezing, screaming for help, slamming their shells shut, baring their teeth, or bristling quills.[1] But in many animals, including humans, fear can be a collective emotion as well as an individual feeling. In social animals the giving of alarms and the communication of fear are of obvious survival value. Many animals respond to signs of danger by alerting others. They give alarms. In an extreme form, the spreading of fear through the group results in panic.

Some alarms are visual. A pigeon that takes off suddenly causes the rest of its group to take alarm and fly off too. The white tails of rabbits and of white-tailed deer are especially conspicuous when they are running, and serve as alarm signals to other members of the group.

Some alarms are smells. For example, a fright-inducing substance under the skin of minnows is released when they are injured, causing other members of the school, and even fish of other species, to avoid going near. Ants alert other members of their group to danger by releasing alarm substances.[2] Some go further: aggressive slave-making species such as *Formica subintegra* employ these smells not only in defense of their own colonies but also to terrorize their victims. A massive discharge of alarm substances sends the ants they are attacking into a panic, enabling the aggressors to take over their nest with little need to fight.[3]

Many species, including blackbirds, have special alarm calls that alert other members of their group to danger. They often alert members of other species too.

Barking is the canine alarm call. Dogs have made themselves useful for tens of thousands of years by warning of people approaching and by alerting their human companions to other dangers. This may well have been their primary function in the early stages of domestication.[4]

Some canine warnings occur when dogs smell, hear, or see a potential source of danger. Some result from an animal picking up intentions, as when dogs warn their owners of a hostile person's threatening intent. And some arise from a dog's ability to pick up intentions at a distance, as in the case of dogs that know when their owners are coming home. Here, of course, signaling their anticipation is not a warning of danger but an announcement.

Dogs and other domesticated animals can warn us of danger by sending out an alarm or by showing obvious signs of fear and distress; and sometimes they actually take practical steps to help or defend us.

## What is epilepsy?

In Leesburg, Virginia, two or three times a week Christine Murray's dog, a Pit Bull and Beagle mix named Annie, leaps onto her lap and licks her face furiously. Christine stops what she is doing, lies down, and a few minutes later is racked by an epileptic fit. "It's amazing," she says, "I can't explain it, I don't know why, but Annie can tell when I'm going to have a seizure."[5]

Annie is not unique. Many other dogs give warnings of epileptic seizures. How do they do it? No one knows. But they make a big difference in the lives of people with epilepsy.

Epileptic seizures, fits, turns, attacks, and blackouts happen when normal brain activity is suddenly disrupted. In the most dramatic type, the grand mal seizure, sufferers become rigid and may fall if standing. They then undergo convulsions, suffer from labored breathing, and may become incontinent. At the beginning of the fit, they may cry out and stop breathing and then their faces may turn blue.

Such seizures can be very alarming to watch, but the person having the seizure is not in pain and usually remembers little of what has happened. After a few minutes, the fit ends spontaneously and the person recovers. He or she may at first be confused and, if the fit has occurred in public, may be embarrassed, especially if he or she has been incontinent. Some sufferers then pass into a trancelike state and behave unpredictably.

Not all kinds of epilepsy involve convulsions, and some seizures affect only part of the body. The mildest form of epilepsy, traditionally called petit mal, involves a brief interruption of consciousness without any other signs, except perhaps for a fluttering of the eyelids. Such fits occur most commonly in children and are usually known as "absences."

Epilepsy, the most common of the serious neurological disorders, can affect people of all ages. About one person in two hundred suffers from it, and in many cases the fits begin in childhood. Although it can often be controlled by drugs and by avoiding situations that bring it on, some people continue to have seizures despite taking appropriate precautions.

In ancient times, epilepsy was known as "the sacred disease." No other disease has aroused so much folklore, probably because none is so suggestive of possession. One of the problems that epileptics have to face is the social stigma attached to the disorder and the uneasiness that many people feel in their presence.

Most people with epilepsy are able to lead reasonably normal lives, although for obvious reasons they are not allowed to drive. For those whose attacks cannot be completely controlled by drugs, one of the greatest difficulties is their unpredictability. In some cases a pattern of preliminary symptoms, known as the aura, which can include uncontrolled twitching in parts of the body, sensations, or bizarre behavior, comes on before the fit begins. But in many cases the disturbance spreads so rapidly that the patient is unconscious before he or she has time to notice anything.

No one wants to be walking along the street, shopping, or climbing stairs when an attack begins. Even within the safety of home there is danger of injury if a blackout occurs when the person is standing up. This is why dogs that know when an attack is about to occur can change the lives of epileptics.

## The predictive behavior of dogs

For many years there have been stories of dogs that anticipate epileptic fits, but almost no research was done on the subject until recently. Most dogs that give warning to their owners do so spontaneously and not as a result of any special training. Hilary Spate of Little Sutton, South

Wirral, described this behavior by her dog: "When I am at home, Penny, my Doberman, seems to predict my epileptic fits and . . . she pushes me into my chair. I mentioned this to a doctor, but he just smiled. I have warnings myself, but Penny always beats me to it. She has never failed me in my home. Outside she stays by me until help arrives."

Ruth Beale, whose Golden Retriever, Chad, won the British Therapy Dog of the Year Award in 1997, had a son who suffered from both petit mal and grand mal seizures. Chad alerted Ruth several minutes before her son had a grand mal seizure, but he usually ignored the lesser attacks. "He will come up and start pawing at my lap to get attention and sometimes he will bark as well." This often occured when Ruth was in a different room from her son. She was able to go to him in time to prevent him from falling or having some other accident.

Some dogs give the warning only a few minutes in advance; others can alert their owners half an hour or more before an attack. Antonia Brown-Griffin of Kent suffered up to twelve major seizures a week and was housebound until she took on a helper dog named Rupert, who became her lifeline to the outside world: "He can sense, up to fifty minutes before, that I am going to have an attack and taps me twice with his paw, giving me time to get somewhere safe. He can also press a button on my phone and bark when it is answered, to get help, and, if he thinks I'm going to have an attack while I am in the bath, he'll pull the plug out. I just can't imagine life without him."[6]

No one knows how many people with epilepsy are fortunate enough to have dogs that warn them, but there are probably many thousands worldwide.

## The pioneering research of Andrew Edney

In the early 1990s a British veterinarian, Andrew Edney, carried out the first systematic survey of the warning behavior of dogs prior to epileptic seizures. By placing an appeal in *Epilepsy Today,* the newsletter of the British Epilepsy Association, and in other journals and newspapers, he contacted epileptics who owned dogs. He studied in detail twenty-one dogs that seemed able to predict attacks. No particular breed, sex, or age predominated in this sample; the seizure-alert dogs included working dogs, sporting dogs, terriers, toy dogs, and mixed breeds, males and females, young dogs and old ones.

On the basis of questionnaires, Edney was able to compile a profile of the dogs' behavior before the seizure began. Typically, they looked anxious, apprehensive, or restless. They alerted people in the vicinity or went elsewhere to seek help. Barking and whining were frequent, as were jumping up, nuzzling the person, and licking the person's hands or face. The dogs sat by their owners or herded them to safety and encouraged them to lie down. While the seizure was taking place, they either stayed beside the person, some licking the face or hands, or went to seek assistance. And they were remarkably reliable. As Edney commented: "No dog seemed to get it wrong—one even ignored 'fake' seizure attempts."

None of the animals in Edney's sample had been trained. They all showed their warning behavior spontaneously. And most of the epileptics had to discover their animals' behavior for themselves. Some of them commented that it took some time before they realized the significance of their dogs' signals.

Edney concluded that the behavior observed before a seizure was largely aimed at getting attention and seemed to be "designed to stop subjects in their tracks so action can be taken," Edney wrote. "Action during the seizure is fairly consistent. It appears to be directed toward protection and resuscitation, as well as . . . alerting others in the proximity."[7]

Recent studies at the Universities of Florida[8] and Calgary[9] have confirmed Edney's findings, and have again shown that seizure-alerting behavior is not confined to any particular breed, age, or gender of dog.

## Cats and rabbits

With two exceptions, all the reports of warning behavior prior to seizures of which I am aware concern dogs. But a few animals of other species also behave in a similar way. The first exception is a rabbit belonging to Karen Cottenham, of East Grinstead, Sussex.

Karen used to sustain terrible injuries when she collapsed during epileptic fits, including broken ribs, fractured ankles, and cuts on her face. She and her husband bought a rabbit, Blackie, and because she did not like to keep him in a hutch outdoors in the cold, she house-trained him and kept him indoors. Soon afterward she realized that he "flitted" around her legs before she had a seizure, enabling her to get to safety. When Blackie died, she bought another rabbit, Smokie, who soon took on Blackie's role: "I don't know how or why, but several minutes before I have a fit, Smokie darts around my legs in a frenzy. I know I have to get to bed or lie on the floor so I won't fall over. When I come 'round, Smokie is usually nestled by my face, as if willing me back to consciousness."[10]

Kate Fallaize, who lives in Staffordshire, has a five-year-old tortoiseshell cat that warns her of impending fits up to an hour in advance: "Before I have a fit she starts to act strangely. She keeps coming right up to my face and staring at me, and she sits next to me and touches me with her paws every few seconds. She will not leave me or let me out of her sight. I now go and lie down if she starts doing this." The cat stays with her throughout the fit and is still there when Kate wakes up. This cat has not been trained to give warnings, and Kate's previous cat did not do so, although he did stand guard over her after the fit had come on.

Epileptic seizures are not confined to humans. Other species have them too, including a dog in Brooklyn, New York, belonging to Karen Wiegand. Karen's cat forewarned her of the dog's seizures several hours in advance, allowing her to be nearby to take care of him: "It took me quite some time to notice that whenever my dog was going to have a seizure, my cat would lay across his body. Once the seizures began, she would go upstairs and steer clear, but she would repeat this each time she anticipated the onset of the seizures. After his seizures were done she would go back to sleeping alone or near him but not on him, until the time was coming near again."

## The training of seizure-alert dogs

The training of dogs to give warnings of seizures is being pioneered in Britain by a small charity in Sheffield, called Support Dogs.[11] The training manager, Val Strong, trained dogs to respond demonstratively to signs of a seizure. The first successful training of this kind was with Molly, a rescued Collie–German Shepherd cross. To start with, Molly was being taught not to be a seizure-alert dog but simply to help her epileptic owner, Lise Margaret. First she was trained to a good level of general obedience. She then learned specialized tasks such as fetching Lise a blanket after a seizure to prevent her from getting too cold, and bringing her the telephone. "It can be difficult to talk, so now I just press a programmed number and Molly will bark into the phone. Friends know that I need assistance."

Molly would have been a big help to Lise even if this had been the limit of her talents, but Val Strong had a hunch that she could go further. She began videotaping Lise and Molly, and after examining many hours of tape noticed a definite but subtle change in Molly's behavior about thirty minutes before Lise had a fit. Molly began staring at Lise. "We just needed to encourage her to be more demonstrative. She's very dramatic now—she barks and licks." Building on this experience, Support Dogs subsequently trained other dogs to alert their owners to oncoming seizures.

In the United States the training of seizure-alert dogs is being coordinated by the National Service Dog Center of the Delta Society.[12] The Delta Society is also helping to raise awareness among the public of service dogs in general. Whereas guide dogs for blind people are widely recognized and are admitted into shops and restaurants where pets are normally not allowed, there is less recognition of hearing dogs for deaf people, assistance dogs for people with disabilities, and seizure-alert dogs. For example, Christine Murray's dog, Annie, was barred from some restaurants and stores in Virginia. "I try to tell them she's a seizure dog," Christine said, "but they don't believe me."[13] As seizure-alert dogs become better known, this problem should diminish.

In a recent study Val Strong, together with several experts on epilepsy, found that most people provided with seizure-alert dogs gained an additional benefit: They had fewer seizures.[14] The dogs' warnings gave them a greater feeling of well-being and confidence, as well as an ability to engage in more activities, which somehow reduced the frequency of seizures.

## How do they know?

So far almost no research has been done on the ability of dogs to predict epileptic fits, and no one knows how they do it. The three most common theories are (1) that the animal notices subtle changes in behavior or muscular tremors of which the person is unaware; (2) that it senses electrical disturbances within the nervous system associated with an impending seizure; and (3) that it smells a distinctive odor given off by the person before an attack.

All three possibilities would require the dog to be quite close to the person. Indeed, the detection of electrical changes in the nervous system, if it's possible at all, would require them to be very close indeed. Dogs could not be expected to react if they are out of range of sight or smell.

However, as we have seen, some dogs seem to react to their owners' thought and intentions at a distance, and some also seem to know when they have had an accident or are dying, even hundreds of miles away. It may therefore be worth considering the additional possibility that the dogs are not simply reacting to subtle sensory cues, but may be picking up signals of a nature as yet unknown to science. Can dogs give warnings of seizures even if their owners are out of sight and some distance away from them?

Normally, epileptics like to be close to their dog at all times, so that it can warn them when a fit is coming on. But I know of three cases in which the dog seems to know even if it is in another room. Steven Beasant, of Grimsby, Lincolnshire, was regularly warned of impending fits by his mixed-breed dog, Jip. Normally Jip followed him around and stayed very close prior to an attack, and when Steven was sitting down, the dog jumped up on him. But Steven said that Jip sometimes "comes bounding through from the kitchen and then he will pin me to the chair." So whatever signals Jip was reacting to could be felt in a different room. The same seemed to be true of Sadie, a Doberman that belonged to Barbara Powell of Wolverhampton. Until she died at age thirteen, Sadie gave warnings of her owner's impending seizures by whining. She usually did this when she was in the same room, but sometimes she still did it in a different room.

Dr. Peter Halama, a neurologist in Hamburg, Germany, had an epileptic patient whose dogs reacted when they were in a different room:

> Before an attack her dogs (two mixed breeds, one male, one female) stay close to her, and as soon as the attack strikes they try to help her.

> One of them even tries to get between her and the floor when she falls. When she lies on the floor they lick her face and hands until she recovers full consciousness. They do not permit any other person to get close to her when she is in this condition. When she is in a different room from the dogs, shortly before an attack they come running to her room and stay with her, ready to help her in the same way. Her husband has often seen it and can testify to this behavior.

Interestingly these dogs also predicted the woman's homecomings. Her husband found that they "get restless and go to the front door before she comes back from shopping (at irregular times). They show this behavior twenty to thirty minutes before her arrival." If these dogs could react telepathically to the woman's intention to return, then perhaps their reactions to impending epileptic attacks when in a different room could also have involved telepathy. But there is a major difference between these two situations. Her coming home involved a conscious intention; but the onset of a fit was neither conscious nor intentional.

## Pets and diabetics

Some dogs belonging to diabetics give warnings when their owner's blood sugar levels are dangerously low. Such a hypoglycemic attack can lead to a coma, an epileptic seizure, and even death.

Alan Harberd, of Chatham, Kent, had a Collie called Sam who warned him when he was hypoglycemic. If this happened when Alan was asleep, Sam woke him before he slipped into a coma. "My blood sugar level is low, but not so low I can't get up and do something about it. It is touch and go sometimes, but it is uncanny the way he does it."

A pioneering survey was published in the journal *Diabetic Medicine* in 1992 by a group of clinicians at the Bristol and Berkley Health Center in Gloucestershire, who interviewed forty-three pet-owning patients who had suffered from hypoglycemia. Fifteen of them said they had noticed reactions on the part of their animals. Fourteen of these animals were dogs. They helped their owners by barking, fetching neighbors, or other appropriate responses.[15]

Some cats give warnings too, and wake their owners in the night when their blood sugar levels are dangerously low.

In 2000 the *British Medical Journal* published a study from the University of Liverpool entitled "Non-invasive detection of hypoglycaemia using a novel, fully biocompatible and patient friendly alarm system," that is, a dog. In one of the case histories, a forty-seven-year-old diabetic woman had up to two hypoglycemic episodes a week, usually in the afternoon or sometimes at night: "Within the past year her seven-year-old mongrel bitch, Susie, has shown peculiar behavior during the patient's hypoglycemic attacks. At night she has been nudged awake by Susie and has found herself to be hypoglycemic. Susie goes back to sleep only after the patient has taken carbohydrates and her symptoms have settled; interestingly her husband sleeps throughout. On other occasions, Susie . . . has prevented the patient leaving the house until she has taken food to correct hypoglycemia."[16]

How do dogs know? The authors of the Liverpool study noted that direct contact with the patient was not necessary; they suggested that possible clues included sweating and muscle tremor, and added, "We are attracted to the notion of the 'sixth sense' with which dogs are commonly credited but acknowledge this will have to be substantiated by further research."[17]

## Diagnosing cancer

Several pet owners say that their animals have helped to diagnose cancer and other ailments, and some cases have been reported in the medical literature. Hywel Williams and Andrew Pembroke of King's College Hospital, London, wrote of a woman who was referred to their clinic with a lesion on her left thigh, which turned out to be a malignant melanoma.

> The patient first became aware of the lesion because her dog (a cross between a Border Collie and a Doberman) would constantly sniff at it. The dog (a bitch) showed no interest in other moles on the patient's body but frequently spent several minutes a day sniffing

> intently at the lesion, even through the patient's trousers. As a consequence, the patient became increasingly suspicious. This ritual continued for several months and culminated in the dog trying to bite off the lesion when the patient wore shorts. This prompted the patient to seek further medical advice. This dog may have saved her owner's life by prompting her to seek treatment while the lesion was still at a thin and curable stage.[18]

There are several similar cases on our database. For example, Joan Hart, of Preston, Lancashire, found that when she sat down with her slippers on, Lady, her Sheltie bitch, would take off one particular slipper and lick her instep. Joan had a cyst there, and she eventually went to her doctor. He thought it was a wart but sent her to the hospital for some tests, to make sure. It turned out that Joan had a rare type of malignant cancer. She said, "I wish I had taken more notice of Lady, who was trying to tell me about it."

Hazel Woodget is convinced that her Chihuahua, Pepe, saved her life. In May 2001 when she was sitting on her sofa, "Pepe started pushing his head under my arm and then pawing my left breast, as if he were trying to dig a hole. I pushed him away but he kept coming back. He stared at me intently and then pressed his front paws on my breast. An excruciating pain shot through me and I leaped up in agony." She went to her doctor and was diagnosed with cancer on the breastbone, just where Pepe had indicated. She had a mastectomy, but a month later Pepe was staring at her breast again: The cancer had returned. On two further occasions Pepe diagnosed a recurrence of cancer. Hazel had a second mastectomy and has now been successfully treated. She says, "I owe him my life."[19]

Fortunately, some medical researchers are now paying more attention to the diagnostic abilities of dogs. In 2004 the *British Medical Journal* published a study by John Church and his colleagues showing that dogs could be trained to detect human bladder cancer by sniffing samples of urine. And in 2006 a study in California showed that dogs could learn to detect lung and breast cancers by sniffing samples of the patients' breath. Under double-blind conditions, trained dogs were tested with exhaled breath samples from cancer patients whom they had not previously encountered and with samples from a similar number of healthy people who served as controls. They identified the samples from patients with cancer with an accuracy rate of 98 percent.[20]

Obviously in these tests with urine and breath samples the dogs were using their sense of smell, but this does not necessarily explain the behavior of Hazel Woodget's dog, Pepe, or other similar cases. Pepe had not been trained to sniff cancer, and in any case Hazel's tumor was inside her body, not on her skin. As in the case of dogs detecting hypoglycemia, smell might provide an adequate explanation some of the time, but dogs may also be using abilities that we do not yet understand.

## Animals that warn of other kinds of illness

Epileptics have seizures repeatedly and therefore have time to recognize and pay attention to any signals their animals may be giving them. But some animals also seem to anticipate other kinds of illness before anyone has noticed the symptoms. Their reactions may well be misunderstood at first. For example, a German Shepherd belonging to the Albrecht family of Limbach, Germany, started to follow the woman of the family, Hilde, for no apparent reason, looking at her in a strange way and whining. "I told my husband to go to the vet with her because something must be wrong. A few weeks later it was I who was ill, not the dog, and I had to have an operation." Several years later the dog behaved the same way with Hilde's daughter, who later turned out to have appendicitis, and the same thing happened with that girl's younger sister.

Similarly, Christine Espeluque, of Nissan-les-Enserune, France, had a Cocker Spaniel that seemed to know in advance when her young children were going to fall ill. With her five-year-old son, she said, "before the illness breaks out [the dog] begins to follow him everywhere. She gets on the chair he is sitting in, sleeps on his bed with him the whole night, cries all day long when he is in school until he comes home. Now that I've gotten used to it I always know ahead when my children will be ill." But the dog only reacted to the children in this way. She did not predict their mother's illnesses.

Sometimes a dog's warning is so unmistakable that it is effective the first time it happens. Esther Allen of Bushbury, in the West Midlands, wrote,

> I was decorating the living room ceiling, and to reach it I had to stand on a chair on the table. I only had about another square foot to finish, when Fara, my miniature longhaired Dachshund, got up on a chair, then on to the table, and started tugging at my skirt. I said, "Just a minute, I won't be long," but she wouldn't let up, so I got down. As I reached the floor I blacked out for a few seconds. When I came to she was licking my face. If Fara had not made me get down off the table and chair, I certainly would have had a serious injury.

The Dachshund's behavior resembles that of the seizure-warning dogs of epileptics. But how did she anticipate her owner's collapse? In a similarly puzzling way, some dogs anticipate heart attacks and take action that minimizes the damage caused by falling down. Hans Schauenburg of Roelbach, Germany, reported, "My partner suffered from heart attacks several times, so that she simply collapsed. Rolf, our German Shepherd, normally quite a rough fellow, always anticipated these attacks and placed himself in front of his mistress in such a way that she never fell on her head but always on her back."

Obviously, the first possibility that needs to be considered is that the animal picks up some subtle change in the person's behavior, movements, or scent. But animals sometimes react when the person is in a different room or farther away. Erni Weber of Grosskut, Austria, told of a cat that reacted even while his person was out of the building:

> One afternoon in July my husband went for his usual walk before supper. Ten minutes later our cat displayed unusual behavior. He ran through the flat restlessly, growled to himself, and had his back hair standing on end. After one hour my husband came back and said, "I don't feel well. I'll lie down a little before the meal." He went into the bedroom, and I continued my work in the kitchen. Suddenly Aimo became even more restless and pushed his muzzle against my legs. Then he ran away from the kitchen, looking behind to see whether I was following him. Like a dog he led me to the bedroom, where I found my husband writhing with kidney pains. We called an

emergency doctor and he relieved my husband's pain. Soon afterward Aimo was our good old quiet cat again.

If this were an isolated case, it would be tempting to suppose either that it was just coincidence or that the cat had noticed some signs before Herr Weber set out on his walk, and that it took Frau Weber some time to notice the cat's unusual behavior. But we have seen that many animals react when their owners die unexpectedly in distant places or when they are in distress. In that context it is not so surprising that a cat should perceive that its owner was not well when he was out for a walk.

## Forebodings of sudden death

The reactions of pets before the onset of illness are easily misunderstood, and their meaning becomes clear only in retrospect. The same is true of unusual behavior prior to sudden deaths.

In 1995, Christine Vickery was living in Sacramento, California, with her husband, whom she describes as a "fitness fanatic, fifty-two years old and very fit." He started each day with vitamin pills, ate a low-fat diet, worked out on a cardioglide exercise machine, and walked part of the way to work.

> On the evening of December 1 he arrived home at 6:30 as usual. Instead of running to greet him, my dogs, Smokie and Popsie, stayed in their baskets in another room. He called them. They refused to move. At 9:00 P.M. the dogs came to the living room and sat at my husband's feet, staring up at him. My husband was upset and wondered what (as he said) they knew that he did not. They kept up this odd ritual for the next five days. On the night of December 6, the older dog, Smokie, caressed my husband's leg with his nose. Popsie offered him a paw. At 1:30 A.M. on December 7 my husband died in his sleep. I envied my dogs. They had known somehow and had said their good-byes.

Cats can have similar presentiments. Dorothy Doherty of Hertfordshire says that the day before her husband collapsed and died, their cat continually rubbed around his legs: "I remember him saying 'What's wrong with her today?' As she had never been so persistent before, I have often wondered if she knew what was going to happen."

There are many other examples of dogs and cats forewarning of medical emergencies and sudden deaths. But like all premonitions, their significance is apparent only in retrospect. Skeptics will say that there must be thousands of cases of unusual behavior not followed by death or disaster and soon forgotten, so there is nothing more mysterious at work than coincidence and selective memory. But although this standard argument may sound scientific, it is no more than an untested hypothesis. Skeptics have not collected any statistics to support it.

The arguments of skeptics are of scientific value if they are treated as reasonable possibilities to be tested; but they are antiscientific if they are used as a way of inhibiting inquiry. Unfortunately, this has all too often been the case. That is why we still know so little about these fascinating phenomena.

*Chapter 15*

# Forebodings of Earthquakes and Other Disasters

On September 26, 1997, a major earthquake devastated the Basilica of St. Francis in Assisi, Italy, and caused much damage in nearby towns and villages. Shortly before the earthquake many people noticed that animals were behaving strangely. The night before, some dogs barked much more than usual; others were strangely agitated and restless. Observers said that cats seemed nervous and disturbed, and some went into hiding. Pigeons flew "strangely." Wild birds fell silent a few minutes before the earthquake struck, and pheasants "screamed in an unusual way."[1]

Some changes in the behavior of animals were noticed several days beforehand. Silvana Cacciaruchi reported that "A friend told [her], 'Don't go to eat at the taverns by the river in Foligno because there are rats along the river, big ones.' At least a week before the earthquake people started saying that Foligno was invaded by rats. I have been living here for a long time, and this never happened before. Rats were everywhere, but nobody connected it to the earthquake."

Foligno is 12 miles from Assisi, and was badly damaged by the quake. Why did the rats leave the sewers? How did so many other animals seem to anticipate the catastrophe?

Skeptics explain away such stories in terms of coincidence and selective memory: people may recall odd behavior only if it is followed by an earthquake or other catastrophe; otherwise they forget about it. No doubt there is some truth in this argument. But it would be rash to dismiss all the evidence in this way. Many experienced observers of animals are convinced that animals have indeed behaved oddly prior to earthquakes. Within three weeks of the Assisi earthquake, while aftershocks were still occurring, Anna Rigano, my Italian research assistant, was on the scene in Assisi, Foligno, and other earthquake-affected areas of Umbria, where she interviewed dozens of people, including pet owners, pet shop proprietors, and veterinarians, while memories were still fresh. Most had noticed unusual behavior in animals prior to the earthquake, and most were confident that this behavior was indeed exceptional.

Similar patterns of animal behavior prior to earthquakes have been reported independently by people all over the world. I cannot believe that they could all have made up such similar stories or that they all suffered from tricks of memory.

The first detailed description from Europe concerns a cataclysmic earthquake in 373 B.C. at Helice, Greece, on the shore of the Gulf of Corinth, which swallowed the city up. Five days before the quake, according to the historian Diodorus Siculus, rats, snakes, weasels, and other animals left the city in droves, to the puzzlement of the human inhabitants.

Other reports from ancient times include the statement of the Roman writer Pliny the Elder that one of the signs of a coming earthquake is "the excitation and terror of animals with no apparent reason." There were similar accounts in the Middle Ages, including one from Württemberg in 1095: "The fowl left human habitations to go and live wild in the woods and mountains."[2] In recent centuries, the strongest earthquake to shake Europe happened in 1755 at Lisbon, Portugal. It resulted in enormous devastation and was so powerful that the earth's motion caused church bells to ring as far away as Sweden. This earthquake was discussed by many contemporary writers, including the philosopher Immanuel Kant, who wrote about the signs of an impending earthquake: "Animals are taken with fright shortly before it. Birds flee into houses, rats and mice crawl out of their holes." There were reports of a "multitude of worms" coming out of the ground eight days before the Lisbon earthquake, and of cattle being "highly excited" a day before.[3]

Hundreds of other examples have been chronicled by historians, as have many more recent cases—for example, "Before the Agadir earthquake in Morocco in 1960, stray animals, including dogs, were seen streaming from the port before the shock that killed 15,000 people.[4] A similar phenomenon was observed three years later before the earthquake which reduced the city of Skopje, Yugoslavia, to rubble. Most animals seemed to have left before the quake."[5] Before the earthquake that destroyed much of Kobe, Japan, on January 17, 1995, unusual behavior was observed in mammals, birds, reptiles, fish, insects, and worms.[6]

Since the first edition of this book was published in 1999 I have continued to monitor major earthquakes to find out if they were preceded by unusual animal behavior. They were. For example, on August 17, 1999, a devastating earthquake struck Turkey, with its epicenter near Izmit. Dogs were howling hours beforehand, and dogs, cats, and birds were behaving strangely. The next month the same happened in Greece several hours before a major earthquake with its epicenter near Athens. On February 28, 2001, there was a 6.8-magnitude earthquake near Seattle. Some cats were said to be hiding for no apparent reason up to twelve hours beforehand, and dogs, goats, and other animals were behaving strangely.[7]

One of the very few systematic observations of animal behavior before, during, and after an earthquake concerns toads in Italy. A British biologist, Rachel Grant, was carrying out a study of mating behavior in toads for her Ph.D. project at San Ruffino Lake in central Italy in the spring of 2009. To her surprise, soon after the beginning of the mating season in late March, the number of male toads in the breeding group suddenly fell, from more than ninety active toads on March 30 to almost none on the following days. As Rachel Grant and her colleague Tim Halliday observed, "This is highly unusual behavior for toads; once toads have appeared to breed, they usually remain active in large numbers at the breeding site until spawning has finished." Six days later, on April 6, Italy was shaken by a 6.4-magnitude earthquake, followed by a series of aftershocks. The toads did not resume their normal breeding behavior for another ten days, two days after the last aftershock. Grant and Halliday looked in detail at the weather records for this entire period but found nothing unusual. They were forced to conclude that the toads were somehow detecting the impending earthquake five to six days in advance.[8]

Yet in spite of this wealth of evidence, most professional earthquake researchers ignore the stories of animal warnings or dismiss them as superstition or selective memory. As far as I know, none of the hundreds of millions of dollars a year currently spent on seismological research in the West are devoted to investigating the reactions of animals. Here is yet another area where taboo and prejudice have closed the minds of professionals and where skepticism serves to inhibit rather than promote scientific inquiry. But it is not just our scientific understanding that is impoverished by this attitude: animals could save lives by providing valuable warnings.

## Predicting earthquakes

Many politicians and taxpayers believe that the large amounts of public money spent on seismological research will help to develop methods of earthquake prediction. But unknown to those who fund them, most professionals believe that detailed predictions are impossible, and they no longer even try to make them. An article by four eminent experts published in the American journal *Science* in 1997 states its thesis succinctly in the title: "Earthquakes Cannot Be Predicted." It quotes with approval the remarks made in 1977 by Richter, developer of the eponymous magnitude scale, who dismisses earthquake prediction as "a happy hunting ground for amateurs, cranks, and outright publicity-seeking fakers." The experts see the role of seismology as contributing to "earthquake hazard mitigation" rather than predicting specific earthquakes: "Statistical estimates of the seismicity expected in a general region on a time scale of thirty to one hundred years and statistical estimates of the expected strong ground motion are important data for designing earthquake-resistant structures. Rapid determination of source parameters (such as location and magnitude) can facilitate relief efforts after large earthquakes."[9]

These are indeed useful roles for seismology to play. But while this cautious attitude protects seismologists against making public mistakes through issuing false alarms, it justifies the continued neglect of research on the warnings that animals can give.

By contrast, in China in the 1970s, earthquake researchers actually encouraged members of the public to watch out for and report possible signals that, according to age-old Chinese traditions, were believed to herald a catastrophic earthquake.

In June 1974, the Chinese State Seismological Bureau issued a warning that a serious earthquake should be expected in Liaoning Province in the next few years, on the basis of a historical analysis and geological measurements. As a consequence, the scientific observation network was expanded, and groups of amateur observers were organized in factories, schools, and agricultural communes. More than 100,000 people were trained to watch out for unusual behavior by animals and changes in the level and cloudiness of water in wells, as well as strange noises and unusual kinds of lightning.

In the middle of December 1974, snakes came out of hibernation, crawled from their burrows, and froze to death on the snow-covered surface. Rats appeared in the open in large groups and were often so confused that they could be caught by hand; cattle and fowl were strangely excited; and the water in the springs became cloudy. There was a minor earthquake on December 22, but throughout January 1995 reports of unusual animal behavior continued, with more than twenty species showing signs of great fear. Plans were made to evacuate Haicheng, a city of half a million people.

At the beginning of February the number of reports climbed steeply when cattle, horses, and pigs began panicking. According to one observer, "Geese flew into trees, dogs barked as if mad, pigs bit each other or dug beneath the fences of their sties, chickens refused to go into their coops, cattle tore their halters and ran away, and rats appeared and acted drunk. . . . Groundwater anomalies began spreading."[10]

On the morning of February 4 the decision was made to evacuate Haicheng. The same day, at 7:36 P.M. the anticipated earthquake finally came, with an intensity of 7.3 on the Richter scale. More than half of the buildings in the city were destroyed. Tens of thousands of people might have lost their lives had it not been for the timely warning. There were victims, but "Most of those were people who had put too little faith in official earthquake predictions to put up with February temperatures outdoors."[11]

For a while, some Western seismologists were impressed. The possibility of using anomalous animal behavior for earthquake warnings was even discussed within the U.S. Geological Service.[12] But within a few years the conventional skepticism predominated again, and the idea was dropped. Nevertheless, the Chinese have continued their earthquake prediction program. They have had some spectacular failures, most notably the unpredicted Tangshan earthquake of 1976, in which at least 240,000 people died. But they have continued to make successful predictions. For example, in 1995, they warned local authorities in Yunnan Province one day before a major earthquake struck in 1995.[13] On April 5, 1997, Xinjiang seismologists predicted that an earthquake between magnitudes 5 and 6 would strike within a week. According to a report in *Science*, "During the night, authorities evacuated 150,000 people to shacks and canvas shelters. Early the next morning, a magnitude 6.4 quake occurred, and at noon a magnitude 6.3 struck. Together they destroyed 2,000 houses and damaged 1,500 more, but no one was killed. Similarly based predictions preceded a magnitude 6.6 quake on April 11 and magnitude 6.3 on April 16."[14]

The Xinjiang scientists did cry wolf once during this period, but over all, the Chinese have been remarkably successful in predicting earthquakes—in striking contrast to their Western counterparts, who do not even try. The Chinese continue to take a practical approach, combining seismological and geological measurements with observations of wells and springs and other "alternative methods"—a euphemism used in Western scientific publications for "unusual animal behavior." But the Chinese seismologists are modest about their achievements and point out that their approach works best when applied to earthquakes with foreshocks, as in Haicheng.[15]

Yanong Pan, a professor at Chaohu College, Anhul Province, was one of the scientists involved in studying the precursors of earthquakes in the 1970s and is convinced that many animals can sense earthquakes in advance. For many years popular rhymes have made millions of Chinese people aware of unusual animal behavior, such as the following, kindly sent to me by Professor Pan:

**震前动物有前兆，密切监视最重要。骡马牛羊不进圈，**

**老鼠成群往外逃。**

**鸡飞上树猪拱圈，鸭鹅怕水狗哀号。冬眠老蛇早出洞，**

**燕雀家鸽不回巢。**

**兔子竖耳跳又撞，游鱼 慌水面漂。家家户户细观察，**

**综合异常早汇报。**

Before the earthquake, animals will have different features or feelings. The most important thing is to keep watching them in case something happens. Horses, donkeys, cows, and goats don't want to go back to their stalls. Mice flee from their homes in groups. The chickens fly onto trees, and pigs try to destroy their sties. Ducks and geese are frightened to go into water, dogs bark in sorrow and pain. Hibernating snakes wake up unusually early. Swallows, pigeons, and other birds fly away. Rabbits have their ears straight up and bounce around, knocking into things. Fish feel they are threatened and are terrified; they stay near the surface of the water without moving. Watch carefully, everyone in each family; observe what's going on, draw your conclusion, and if you are sure tell the government as soon as possible.

In 2010 Professor Pan informed me that the Chinese authorities no longer enlist ordinary people in earthquake prediction. Only official observers are now involved. This is unfortunate: China is in a good position to lead the world in the development of early warning systems through mass participation.

## Research with animals in California

Currently, no official research on the anticipation of earthquakes by animals is going on in the West, as far as I know. Animal behaviorists ignore the subject, and so do seismologists, who concentrate on physical measurements with instruments. Given the successes of the Chinese, this seems like a remarkable omission.

My colleague David Jay Brown and I carried out surveys in California in the 1990s to find out more about unusual animal behavior before two of the most destructive recent earthquakes, the Loma Prieta earthquake of October 17, 1989, which caused much damage in Santa Cruz, Silicon Valley, and other parts of northern California; and the Northridge quake of January 17, 1994, with its epicenter in the San Fernando Valley, in the suburbs of Los Angeles.

Many people in both areas did in fact notice strange and seemingly inexplicable behavior in domestic and wild animals. Said Renata McKinstry of San Jose, "My Cocker Spaniel was really fearful. Her eyes were big, and she ran around and around like crazy, back and forth and back and forth. She would come to me and go away and come to me and go away, like she was trying to tell me, 'You have to get out too.' I thought, 'This dog has gone crazy,' and I was really angry. After about an hour the earthquake happened."

Most of the accounts we have received are about dogs and cats, which may simply reflect the fact that these are the most common pets. Dogs were said to bark for no apparent reason and to snarl, howl, whine, run around, hide, and show signs of nervousness, restlessness, and agitation. Cats seemed nervous or disturbed, and many ran outside or hid. But other animals reacted too. Caged birds became very restless. Horses ran around in unusual ways, goats became agitated, and some chickens stopped laying eggs. Some people noticed that just before the earthquakes struck, there was a strange silence as wild birds and crickets stopped singing.

Some animals showed fear and agitation several days before the Loma Prieta and Northridge earthquakes. Others reacted hours before and still others only a few minutes.

Even emus responded. These giant birds, which are related to ostriches, have a habit of pacing alongside fences. In their native Australia they are often called "birds in search of a fence" because even when they have huge open spaces to range in they will go to a fence and pace alongside it. In Sandy Scott's emu farm in Auburn, Washington, the birds normally walk beside the fences and bed down in their sheds about half an hour before dark. But on two particular evenings their behavior was different: "They were almost running up and down the fence. And when it started to get dark, and they finally did bed down, they bedded down outside their sheds, instead of inside them." On both these occasions, there was an earthquake in the night, several hours after the emus began to show their unusual behavior.

There were also many animals that did not become agitated before the earthquakes struck. Susan Gray of Reseda, near Northridge, California, commented: "The cats were as shaken as we were. It was early in the morning, and both cats were in the bedroom with us. They were down the hall and out the cat door within seconds of the quake beginning." These cats, like many other cats, were terrified and would not come back into the house for days.

Some people, like animals, react before tremors, describing symptoms such as agitation, headaches, and nervousness for no apparent reason. Some say they wake up just before earthquakes; others suffer from unaccountable sleeplessness. Some, like Barry Cane, are especially sensitive to aftershocks: "I could often feel an aftershock coming. It was like there was a change in the atmosphere. I don't really know the words for it, but I'd say, 'Uh-oh, here we go.' And in anywhere from one to five minutes—*boom!*—it would hit."

## An animal-based earthquake warning system

Imagine what could happen in California and other parts of the Western world if, instead of ignoring the warnings given by animals, people took them seriously.

Through the media, millions of pet owners could be informed about the kinds of behavior their pets and other animals might show if an earthquake were imminent. If they noticed these signs, they would immediately call a telephone hot line with a memorable number—say, 1-800-PET QUAKE. Or they could send a message on the Internet. A computer system would then analyze the place of origin of the incoming calls. No doubt there would be a stream of false alarms from people who had misunderstood their pet's symptoms—the animal might simply have been sick, for example—and there might well also be some hoax calls. But a sudden spate of calls from a particular region might indicate that an earthquake in that region was imminent. It would then be important to check that the calls were not due to other circumstances known to affect the behavior of animals, such as dramatic changes in the weather, fireworks displays, a nearby fire, or an influx of predators.

At first this system would have to be used for research purposes only, to see if it worked reasonably reliably. It would not be appropriate to issue any warnings until this had been established. False alarms could cause panic and disruption, and could set back research on this subject for years. Ideally, the reports of unusual animal behavior would be combined with the monitoring of other precursors of earthquakes, including seismological measurements, as in China.

Research in California already indicates that such a system could work. In the late 1970s, following the successful prediction of the Haicheng earthquake, the U.S. Geological Service funded a pilot project based at the Stanford Research Institute. The coordinators, Leon Otis and William Kautz, recruited 1,200 volunteer observers located in earthquake-prone parts of California who undertook to call a toll-free hot line whenever they observed odd animal behavior whose cause was unknown.

This project ran from 1979 to 1981. During this period there were no earthquakes with a magnitude greater than 5 in the areas under observation. Altogether, thirteen earthquakes of 4 and 5 magnitude were appropriate candidates for analysis, although none of them occurred within the areas where observers were concentrated. Seven of these earthquakes were preceded by a statistically significant increase in calls about unusual animal behavior.[16] In some cases the statistics were very impressive indeed.[17] At this stage, however, funding was discontinued and no further research was carried out.

If instead of a mere 1,200 observers, millions could be recruited, a much more detailed assessment of the potential of animal warnings could be made. Pet owners—especially retired people, who have more time and opportunity to observe their animals than people who are out at work all day—could play a vital role in this process.

## How do they know?

Except for some recent experiments in Japan, I am aware of almost no research anywhere in the world on the means by which animals sense an imminent earthquake. There are, however, several theories.

One theory is that the animals pick up subtle sounds, vibrations, or movements of the earth. There are three problems with this theory. First, some of the animals that respond in advance to earthquakes have hearing that is no more sensitive than our own.[18] Second, small earth tremors and minor earthquakes are common in seismically active areas. In 1980, for example, there were 350 earthquakes (excluding aftershocks) of magnitude 3 or less in California.[19] If animals were exquisitely sensitive to weak vibrations, they would give false alarms frequently. They should also respond with fear and alarm to vibrations caused by passing trucks. And third, if so many species of animals can pick up characteristic vibrations before major earthquakes, then seismologists should also be able to identify them with their very sensitive instruments. But they have so far failed to do so despite years of intensive research.

A second theory is that animals respond to gases released by the earth prior to an earthquake. Although some species, such as dogs, are far more sensitive to smells than we are, others, like songbirds, are less sensitive. There seems to be no correlation between animals' sense of smell and their sensitivity to earthquakes. Also there is no evidence that earthquakes are generally preceded by the leaking of gases out of the earth. And if such gases are released through tiny cracks in the earth's surface before earthquakes, then why do animals not respond with fear and panic when people dig holes or mines and when animals burrow?

According to a third theory, animals respond to electrical changes that precede earthquakes—a far more plausible idea than the other two. There is evidence that some earthquakes are indeed preceded by changes in electrostatic fields, which probably result from changes in seismic stress in rocks. It is well known that in some crystals and rocks, changes in pressure generate electrical charges (the piezoelectric effect), and such electrical effects preceding earthquakes could help to explain not only the reactions of animals but also other electrical anomalies, like interference in radio and television broadcasts and strange auras and lights coming from the earth (technically known as seismoatmospheric luminescence).[20]

Conventional seismologists are skeptical about these electrical precursors of earthquakes, but a maverick group in Greece, led by P. Varotsos, claims to be able to predict earthquakes on the basis of geoelectrical signals.[21] And in California the Time Research Institute, headed by Marsha Adams, issues regular earthquake forecasts on the basis of a network of electromagnetic sensors, the input from which is analyzed by specialized computer software.[22]

Meanwhile, Motoji Ikeya and his colleagues at Osaka University in Japan have recently carried out laboratory experiments in which they exposed a variety of animals, including minnows, catfish, eels, and earthworms, to weak electrical currents. Fish showed panic reactions, and earthworms moved out of the soil and swarmed when the current was applied.[23] These preliminary findings could help to explain the anomalous behavior of animals in water and in moist environments before earthquakes. But what about animals that are inside buildings, like dogs and cats? Are they responding to electrically charged ions in the air? Many questions remain unanswered, but this is clearly a promising line of research.

Finally, a fourth theory postulates that animals may sense what is about to happen in a way that lies beyond current scientific understanding. In other words they may be *prescient*, having a feeling that something is about to happen, or *precognitive*, knowing in advance what is going to happen.

This hypothesis would be unnecessary if all the facts could be explained satisfactorily by more conventional theories. Many scientists, myself included, would prefer not to have to consider the idea of influences working backward in time, from the future to the present. I confess that I would prefer to put this idea aside unless I'm compelled to take it seriously. At present, the electrical theory seems sufficiently promising to justify ignoring this more radical possibility.

The trouble is that there are other kinds of animal premonitions that cannot be explained electrically, as we will shortly see. Whether we like it or not, precognition does seem to occur. And if it occurs in other situations, perhaps it also plays a part in premonitions of earthquakes.

## Warnings of tsunamis

On December 26, 2004, one of the most powerful earthquakes ever recorded, magnitude 9.1, thrust up the bed of the ocean off the west coast of Sumatra. It triggered off a series of devastating tsunamis around the Indian Ocean, killing more than 200,000 people and inundating coastal communities with waves of up to 100 feet high.

However, many animals escaped. Elephants in Sri Lanka and Sumatra moved to high ground before the giant waves struck; they did the same in Thailand, trumpeting before they did so. According to a villager in Bang Koey, Thailand, buffalo were grazing by the beach when they "suddenly lifted their heads and looked out to sea, ears standing upright." They turned and stampeded up the hill, followed by bewildered villagers, whose lives were thereby saved. At Ao Sane beach, near Phuket, dogs ran up to the hilltops, and at Galle in Sri Lanka, dog owners were puzzled by the fact that their animals refused to go for their usual morning walks on the beach. In Cuddalore District in South India, buffalo, goats, and dogs escaped, and so did a nesting colony of flamingos that flew to higher ground. In the Andaman Islands, groups of tribal people moved away from the coast before the disaster, alerted by the behavior of animals.

How did they know? The usual speculation is that the animals picked up tremors caused by the undersea earthquake. This is unconvincing.

There would have been tremors all around the Indian Ocean, not just in the afflicted coastal areas. And if animals can predict earthquakes and tsunamis by sensing slight tremors, why can't seismologists?

As in the case of earthquakes, it may well be possible to set up a tsunami warning system by encouraging people in coastal areas to look out for unusual animal behavior and to send in their observations to an easy-to-remember telephone number or to a website. Mobile telephones are now so widespread that even in remote areas people could easily get in touch with a tsunami warning center. Such a warning system would be relatively inexpensive. But national and international tsunami warning programs are still ignoring animals. Once again, prejudice and taboo limit the effectiveness of science.

## Forebodings of storms

"It was a beautiful warm summer day, with a clear blue sky," according to Louise Forstinger of Graz, Austria. "I set off for a long walk with my German Shepherd, Rolly. After we had gone for about one hour he would not go any farther. I tried to make him move on, but nothing helped. I wondered what was wrong. Finally he lay down in the ditch. What else could I do but turn around and go home? Half an hour later the sky darkened and the first thunder could be heard far away. We went on a little faster, and when we had entered the house there was torrential rainfall with hailstones. Then I realized that Rolly must have sensed this much earlier."

Some animals are terrified of thunderstorms and show signs of distress long before their owners are aware of a storm approaching. Dogs and cats often hide. Many other kinds of animals, including horses, parakeets, and tortoises, become apprehensive before storms.

Most of the accounts I have received concern reactions half an hour to an hour before a storm breaks, but in some cases the animal's anticipation begins three hours or more in advance.

The reactions of some animals before storms and before earthquakes are similar, and any animal-based earthquake warning system would need to take this fact into account. Otherwise impending storms could be mistaken for impending earthquakes, resulting in false alarms.

Lightning is of course an electrical phenomenon, and it could well be that some, if not all, of the anticipatory reactions of animals depend on their sensitivity to the electrical changes that precede thunderstorms. This would support the electrical theory of earthquake anticipation. And perhaps some animals with hearing more sensitive than our own hear the thunder when it is still far away. But other animal premonitions cannot be explained in these ways.

## Forewarnings of avalanches

On February 23, 1999, an avalanche devastated the village of Galtur, Tyrol, killing dozens of people. It was the worst avalanche disaster in Austria since 1954. The previous day, the chamois, small goat-like antelopes, came down from the mountains into the valleys, something they never usually do. "The mood in the village became distinctly uneasy. That evening the assistant manageress of his hotel started talking about avalanches in the village, including one thirteen years before that had destroyed a house. The following day, when the first really big avalanche hit, she lost three members of her family."[24]

Albert Ernest worked for nearly fifty years as an avalanche protection officer in the Swiss Alps, mostly in the Enns valley, and is well acquainted with the habits of mountain animals. "Again and again I observed that the chamois were not staying in the danger zone of avalanche breakoffs," he said. "Based on my observations I hold the opinion that wild mountain animals have a presentiment of unsteady situations in the snow cover through an inborn instinct and behave accordingly."

In surveys in villages in the Austrian and Swiss Alps kindly conducted on my behalf by Theodore Itten, the animals most often said to anticipate avalanches were chamois, ibex, and dogs. Some dogs barked persistently for hours before an avalanche struck, and some refused to go outside. Josef Flollriol of Stuben, Tyrol, who had a trained avalanche search dog, said that one morning in March 1988 the dog simply refused to leave the house for his usual morning walk: "We tried several times to get him out, and after thirty minutes a huge avalanche came down beside our house. We would have been dead if we had been outside."

As in the case of earthquakes, it is not clear how these animals anticipated the coming disasters. Perhaps they reacted to electrical or other physical changes. But if so, no one knows what these changes are. Or perhaps they have a more mysterious presentiment of danger.

However they do it, an ability to anticipate avalanches would be of obvious survival value in mountain animals and would be favored by natural selection. But many animals also anticipate man-made catastrophes that would not have occurred in the natural world, such as air raids.

## Warnings of air raids

Teddy Pugh of Birmingham wrote that "During the war, when the German bombing raids were going on, we had a black mongrel dog who used to go to our back door and bark to go out, and you could bet that around ten minutes later the sirens would sound for an air raid. We got so used to the dog doing it that I would run up and down both sides of the street and knock on all the doors to warn of an impending raid. She was never wrong once."

I have received thirty-five other accounts of dogs that gave warnings of air raids before the sirens sounded the official alarm. Some of them let their owners know by whining, some barked, some hid, and others led the way to the air-raid shelter or cellar where the family took refuge. British dogs gave warnings of German air raids during the Second World War, and German dogs gave warnings of British air raids.

Some dogs were said to alert their owners just a few minutes before the sirens went off, but most reacted ten to thirty minutes beforehand. In three cases the dogs gave their warnings more than an hour in advance.

One little dog called Dee sometimes stayed curled up in her basket when the siren went off, and invariably no planes came over. Sometimes when there was no warning siren, she would become very agitated and urge everyone to take cover, and sure enough a raid would take place.[25] As a bonus, some families were able to get back to bed before the all clear sounded: "The dog would suddenly get up, leave the shelter, and settle into its basket with a contented sigh. Five minutes later the all clear would sound."[26]

During the Second World War, cats also anticipated bombing attacks, usually by showing obvious signs of agitation or by hiding. Some were said to give warning more than an hour in advance. Birds too seemed to know when bombers were on their way: seagulls flew off, cock pheasants gave warning calls, and ducks and geese raised the alarm. Here is how a German parrot did it, according to Dagmar Kessel:

> During the wartime year of 1943 I stayed with acquaintances in Leipzig. They had an old parrot. Suddenly, about 9:00 P.M., it was extremely upset in its cage, lifted its left wing and called, *"Da oben! Da oben!"* ("Up there!") It even looked up and nobody could get it quiet. I was surprised and asked my hosts what all this meant. "He always does this before an air alert," the lady said, "usually two hours in advance." That same night the Tommies really came. They destroyed the Crystal Palace.

Warnings given by German pigeons were a source of trouble to an unfortunate Austrian sculptor, Heinz Peteri, who was arrested during the war for his "undiplomatic" words and deported to Bochum, in the Ruhr, to defuse unexploded bombs. He lived in a small room in the tower of the police administration building. From his window he used to watch the pigeons that lived on the roofs, and he noticed that "the birds often flew away suddenly, all of them, and half an hour later (at the most) the bombers came. Afterward the birds came back. This was repeated many times." He used this knowledge to warn his comrades and superiors of impending raids, and his predictions repeatedly proved to be accurate. When the Gestapo heard about it, he was arrested once again under suspicion of being a spy "in contact with the enemy."[27]

During the Kosovo Conflict in 1999, when NATO was bombing Serbia, the animals in Belgrade Zoo provided an early warning about half an hour before the bombs fell. "It's one of the strangest and most disturbing concerts you can hear anywhere," said Vuk Bojovic, the director of the zoo. "It builds up in intensity as the planes approach . . . and when the bombs start falling it's like a choir of the insane. Peacocks screaming, wolves howling, dogs barking, chimpanzees rattling their cages."[28]

How did all these animals know when air raids were imminent? The most obvious possibility is that they heard the enemy planes when they were still too far away for humans to hear. But a few moments' reflection shows that this is not a very plausible suggestion, for at least four reasons. First, as we have seen, the hearing of dogs and other domestic animals is not much more sensitive than our own, although dogs can hear higher-pitched sounds than we can. The bombers used in the Second World War flew at about 250 miles per hour when loaded; hence an animal that responded half an hour before an air raid would have to have heard them when they were about 125 miles away. Some animals were said to respond even earlier, when the bombers would have been more than 200 miles away. Even animals that responded only a few minutes before the sirens went off would have to have heard the planes when they were more than 30 miles away, assuming that the siren gave about five minutes' warning. It is very unlikely that they could have heard the enemy aircraft at such distances.

Second, hearing distant sounds depends on the wind direction, and there is no evidence that the regular warnings given by animals occurred only when the enemy aircraft were upwind. Rather, the evidence suggests that animal warnings were remarkably reliable and not dependent on the wind direction. Moreover, since the prevailing winds in Britain are southwesterly and the German bombers approached from the east, in most raids they would not have been upwind, and hence the sounds would have been blown away from, rather than toward, the animals that sensed their approach.

Third, there were many other aircraft in the skies, including the country's own bombers heading toward enemy territory. Apparently the animals did not warn of the approach of friendly bombers. Therefore the hearing theory would require the animals to distinguish between the sounds of different kinds of bombers at great distances, irrespective of the wind direction. There is no evidence that this is possible.

Finally, during the last year of the Second World War, the Germans were firing supersonic V2 rockets at London. These missiles were launched from Holland and headed upward at about 45 degrees. Their engines cut out after a minute or so, and they followed a ballistic trajectory, reaching speeds of more than 2,000 miles per hour as they plunged downward, arriving unseen and unheard. They took only five minutes to reach their targets in England, some 200 miles away, carrying a ton of high explosives.[29] They were particularly terrifying because their explosion was preceded by no warning and they could strike anywhere in southeast England at any time of day or night.

Dr. Roy Willis, who was seventeen at the time, was living in Essex, just east of London. "I noticed that our dog [a German Shepherd–Norwegian Elkhound cross] was seemingly able to sense the imminent arrival of a V2 rocket. The dog, called Smoke, would go to the window and stare out, hackles raised, as if in anger and fear. After about two minutes, during which time he remained in the same aggressive posture at the window, I would hear the ominous crump of an exploded rocket." At least one other dog owner had a very similar experience, his animal reacting shortly before the explosions. Assuming that these accounts are reliable, and I have no reason to doubt that they are, the dogs could not have heard these missiles coming, however acute their hearing, precisely because they were both silent and supersonic.

If animals were not anticipating air raids by hearing the approaching bombers or rockets, how did they know the attacks were coming?

No explanation is possible in terms of electrical charges in the earth and the atmosphere, such as those that may serve as warning signs before earthquakes. As far as I can see, only two possibilities remain: telepathy and precognition.

**Telepathy.** The animals picked up influences telepathically from people or animals along the flight path of the bombers. A wave of alertness and alarm spread through the human and animal populations as the bombers flew by, and this alarm spread telepathically. The trouble is that this telepathic alerting might have taken place in all directions and hence caused false alarms in places to which the planes were not flying.

Alternatively the animals might have picked up the hostile intentions of the German bomber crews as they moved toward their targets with their attention focused on the places they were planning to attack.

These possibilities are obviously highly speculative, and although telepathy may account for some of the available facts, it cannot account for all of them. In particular, no telepathy theory could explain how dogs could anticipate the arrival of the supersonic V2 missiles: no one was aware of their flight path, and they were unmanned. Even the Germans who launched them did not know exactly where they would land.

**Precognition.** Perhaps the animals somehow intuited what was going to happen in the near future, or at least had an apprehension that *something* was going to happen, without knowing what. This theory would account for the dogs that anticipated the V2 attacks, as well as many other premonitions. One trouble here is that this is a very vague theory. Another is that it raises terrible logical problems and mind-twisting paradoxes, since it implies that something in the future can have an effect backward in time. There is a further logical problem with precognition. It is not possible to know if a precognition is valid until the foreseen event has actually happened. Only in retrospect can precognition be identified as such.

I would prefer to avoid this theory, if possible. I find telepathy easier to accept than precognition. And the two V2 cases are the only evidence so far that seem to necessitate a theory of this kind. However, we must now turn to the many other examples that make the idea of precognition or presentiment almost unavoidable.

## Other kinds of premonitions

In addition to all these examples of warnings that animals gave before air raids, earthquakes, avalanches, and tsunamis, I have received dozens of accounts of apprehensive behavior prior to accidents, catastrophes, and danger.

Some dogs refused to walk along paths when shortly afterward branches or trees fell where the person and the dog would have been. Other dogs, horses, and cats delayed or prevented their owners from setting off on foot or by car when road accidents happened soon afterward, in which they might well have been injured or killed. When one dog adamantly refused to enter a pedestrian underpass, the person with it had no option but to turn back: "We had barely turned around when there was a great bang and the concrete ceiling came down!" Another dog prevented its owner from getting onto a boat that exploded shortly afterward. Another dog pulled its owner away from the roadside just before a van hurtled around the corner and crashed into the place they would have been.

In some of these cases, it is possible, though implausible, that the animals heard something unusual that caused their alarm. In others that is impossible, because the animal's apprehension began long before it could possibly have heard anything that might have given any clue. For example, a woman who was driving with her cat on the back seat of the car, where it normally slept, found that the animal became more and more disturbed. She tried to calm it, but it eventually jumped into the front seat, then went so far as to touch her arm and to slightly bite the hand that held the steering wheel. "So I finally stopped," Adele Holzer reported. "Right at that moment a big tree fell onto the road a few meters in front of the car. Had I continued as before, it would have fallen on the car."

Some of the dangers to which animals alert people are silent, so hearing could have played no part in arousing their apprehension. An Austrian couple were traveling along a steep mountain road with rocks on one side and an abyss on the other when their Poodle, Susi, suddenly started to howl. Friedel Ehlenbeck wrote that "She even put her paws on my husband's shoulders to stop him. My attempts to keep her quiet failed. Her behavior became mad. Startled, my husband slowed down, and when we turned around the next bend we were shocked: the road was gone. Only a few meters in front of us there was a precipice. A landslide had taken the road with it. Susi had saved our lives."

In most cases I have heard of, the behavior of the animals helped protect their people from danger. But not everyone heeded the warnings the animals tried to give. This report came from Elizabeth Powell of Powys, Wales:

> One morning my dog, Toby, tried to stop me going out of the front door. He barged against me, leaned on the door, jumped up at me, and pushed me. He is normally a quiet, loving dog and knows my routine; I would have been back within four hours. I had to lock him in the kitchen and left him howling, something he has never done before or since. I set off at 7:30 A.M. and by 9:40 A.M. I was involved in a horrific traffic accident resulting in a fractured neck and right arm, and many other injuries. When I was in hospital an image of Toby kept

> appearing to me through the drugs, and I could feel his anguish. I sent a mental "Okay, I'll be back soon" and the images disappeared. . . . My husband said Toby was very agitated for twenty-four hours and then suddenly became quiet. I am slowly recovering. In the future I'll listen to Toby.

Sometimes the animal's reactions are not specific warnings about which the person can do anything, but seem to be presentiments of something alarming about to happen. In 1992, Natalie Polinario was living in North London near Staples Corner, where IRA terrorists detonated a large bomb on April 11. Her white German Shepherd, Foxy, was outside in the garden. "I was lying on the bed watching TV. About one minute or two minutes before the bomb went off she came in running and literally crying, in a really weird mood. She got on the bed and just lay there next to me, really stiff as if something had really scared her, but there was nothing out there. And then I heard this almighty bang, which was the bomb at Staples Corner. The minute it went off she was fine again. She has never done anything like it since then, or before."

One night Kerry Greenwood of Footscray, Australia, was suddenly awakened by her cats at 3:05 A.M. "They were terrified, clawing to get under any cover, and shivering and crying. We got up to look around for an invading predator or gas leak, but the house was quiet, so we went back to bed to hug the cats and try to get back to sleep. The cats would not settle but kept burrowing and scratching. They had never done this before or since. At 3:35 A.M. a bomb went off in the Serbian travel agency at the end of the street and blew a huge crater in the road, sending a shock wave that belted through the house, shook all the windows, and scared the hell out of all of us. We got up to look at the fire trucks coming and were so shaken that we sat for the rest of the night in the kitchen. . . . The cats, however, went calmly to sleep as though nothing had happened. I admit I found this comforting; clearly the bomber had done his stuff."

It is hard to avoid the conclusion that some of these warnings must indeed have been precognitive. What other explanation could there be? And if premonitions of disasters, accidents, and air raids can be precognitive, then so might some premonitions of storms and earthquakes, avalanches, and tsunamis, even though others might be explicable in terms of a sensitivity to electrical changes or other physical causes. Some of the forewarnings of epileptic fits, comas, and sudden deaths discussed in the previous chapter might also include an element of precognition.

## Human precognition

All around the world we find people who believe in the ability of some human beings to foresee the future. Shamans, seers, prophets, oracles, or soothsayers are found in most, if not all, traditional societies, and even in modern industrial societies, fortune-tellers and clairvoyants still flourish. No doubt some of them are fraudulent. But there are far too many convincing examples of human premonition to dismiss this entire area of experience out of hand.

Many people who are not professional fortune-tellers have had premonitions that have turned out to be true, and there are many stories of people whose lives have been saved by dreams, presentiments, or forebodings that led them not to take planes that later crashed and to avoid places that would have exposed them to grave but otherwise unexpected dangers. Sometimes they do not or cannot act on these premonitions, either because they are not specific enough or because they do not take them seriously. But sometimes they do.

These different reactions were illustrated in a dramatic way prior to the assassination of President Abraham Lincoln in 1865. A week before he was shot in the Ford Theater in Washington, he told his wife and his friend Ward H. Lamon of a dream in which he heard sounds of mourning in the White House. Anxious to discover the cause, he went from room to room until in the East Room, with "sickening surprise" he saw a catafalque on which rested a corpse in funeral vestments, guarded by soldiers and surrounded by a throng of mournful people. As the face of the corpse was covered, he asked who it was. "The president," he was told, who had been killed by an assassin.[30]

Less well known is the fact that General Ulysses S. Grant and his wife, Julia, were supposed to accompany President Lincoln to the theater and sit in his box. That morning Mrs. Grant felt a great sense of urgency that she, her husband, and their child should leave Washington and return to their home in New Jersey. The general could not leave because he had appointments throughout the day, but as Mrs. Grant's sense of urgency increased, she kept sending him messages begging him to leave. So great was her insistence that he finally agreed to go with her, even though they were scheduled to accompany the president to the theater. When they reached Philadelphia, they heard about the assassination, and later they learned that they were on the assassin's list of intended victims.[31]

Of course, not all premonitions are as dramatic as this, nor do they necessarily involve danger. And many pass unnoticed, especially when they occur in dreams. Precognitive dreams are surprisingly common. A classic book on this subject, *An Experiment with Time* by J. W. Dunne, contains simple instructions for investigating one's own dreams.[32]

Laboratory experiments by parapsychologists have also yielded some impressive evidence for presentiment. In these experiments, people were shown a series of pictures on a computer screen. Most of the images were emotionally calming photographs of landscapes, nature scenes, and cheerful people, but some were emotionally arousing pornographic pictures and pictures of corpses. In each trial, the computer screen was blank to start with. Then one of these images, calm or emotional, appeared on the screen for three seconds. The screen then went blank again. The sequence in which the pictures were shown was randomly determined by the computer. While these tests were going on, the participants' blood pressure, skin resistance, and blood volume in the fingertips were monitored. When people were emotionally aroused, all these changed and provided an objective measurement of their reactions.

Not surprisingly, there were dramatic changes in all these measures of arousal after the emotional images were shown, and these changes did not occur with the calm images. The remarkable feature of the results is that the arousal began *before* the emotional images appeared on the screen, even though nobody could have known by any normal means which picture was coming next. This anticipation began about four seconds before the emotional pictures appeared. These results were highly significant statistically and have been replicated independently in several different laboratories.[33]

These remarkable experiments seem to show that even under laboratory conditions there can be presentiments that something emotionally arousing is about to happen, even though this could not have been known by any normal means.

I believe that we stand on the threshold of a new phase of science, of which this kind of research is just one example. Open-minded inquiry into spontaneous human experience, complemented by laboratory research, can help deepen our understanding of human nature.[34] Further research on the unexplained powers of nonhuman animals can help us to place this understanding in a wider biological and evolutionary context. And premonitions and precognitions may be able to tell us something very important not only about the nature of life and mind but also about the nature of time.

# Part VII

# Conclusions

*Chapter 16*

# Animal Powers and the Human Mind

Most kinds of perceptiveness found in animals also occur in modern people, but to a lesser degree. Why are we so insensitive? Is it because we are human? Perhaps our sensitivity diminished over tens of thousands of years as our brains evolved. Or perhaps the evolution of language has led to a decline in our ability to communicate telepathically, to have premonitions, and to find our way around in unfamiliar places. If so, since all human cultures have language, we would expect human beings in all parts of the world to be less perceptive than animals.

But perhaps this decline in sensitivity is not a feature of our being human or using language, but a more recent phenomenon, a result of civilization, literacy, mechanistic attitudes, and dependence on technology. There seems little doubt that people in traditional nonindustrial communities are often more perceptive than educated people in modern industrial societies.

Many explorers and travelers have reported that telepathic communication and a sense of direction are well developed among the Aborigines in Australia, the Bushmen of the Kalahari, and other hunter-gatherer societies. In Europe, unexplained forms of perceptiveness were generally recognized, like the second sight of Scottish Highlanders[1] and the ability of people in rural Norway to anticipate an arrival by hearing the *vardøger* of a person on his way home. In non-Western civilizations, like India, such forms of perceptiveness are widely taken for granted.

Even in modern societies, there may be differences in perceptiveness between different kinds of people: on average, children may be more sensitive to telepathic influences than adults, and women more than men.[2] On the other hand, men may be more sensitive than women in their sense of direction.[3]

Human and nonhuman perceptiveness do not, of course, exist in isolation from each other. People and domesticated animals have lived together for many thousands of years. People relied on dogs' warnings long before the invention of agriculture. And even before the domestication of dogs, countless generations of our hunter-gatherer ancestors survived by paying close attention to the behavior of wild animals.

A symbiosis has developed between human and animal perceptiveness, and our ancestors may have made up for any deficiencies in their own sensitivity by relying on that of the animals around them. We can still do so today.

## Animal perceptiveness and psychic research

Curiously enough, the unexplained perceptiveness of animals has been ignored not only by mainstream scientists but also by most psychic researchers and parapsychologists.[4] Why?

The main reason seems to be historical. The scientific investigation of telepathy and other psychic phenomena began in the late nineteenth century, when the pioneers of psychic research hoped to investigate scientifically the conscious survival of bodily death. Telepathy was of interest for the light it shed on the nature of the human soul. In this context, psychic phenomena were seen as peculiarly human rather than as part of our biological heritage.

The Society for Psychical Research was founded in Britain in 1882 "to examine without prejudice or prepossession and in a scientific spirit those faculties of man, real or supposed, which appear to be inexplicable on any generally recognized hypothesis." There is nothing here to deny the existence of such faculties in nonhuman animals. But the focus is explicitly on the "faculties of man." The same human-centeredness characterizes parapsychology.

Psychic research and parapsychology are usually regarded as of no importance or, at best, of marginal significance by mainstream scientists. The situation changes radically, however, if telepathy and other unexplained faculties are seen not as specifically human but as part of our biological nature. Then we can recognize that human telepathy is rooted in the bonds that coordinate members of animal societies. The human sense of direction is derived from the ability of animals to find their way home after foraging and exploring. And human premonitions are closely related to forebodings in many other species. Psychic research and parapsychology can at last be linked up with biology, and the phenomena they study can be seen in an evolutionary perspective.

## The power of intention

Human intentions can bring about effects at a distance in a variety of ways: a dog can pick up its owner's intention to come home from many miles away; a cat can respond to its owner's silent call; and a person can feel the intention of someone to call by telephone. Likewise, animals' intentions can affect people to whom they are bonded, as when cats in distress call their owners to the rescue. And animals' intentions can also affect other animals. All these kinds of intentions can work telepathically through morphic fields.

But what if an animal's intentions are directed toward an inanimate object rather than a member of its social group? If its intentions could influence such an object at a distance, without any known forms of physical contact, then this would be an example of *psychokinesis*, the name given by parapsychologists to the action of mind on matter.

In some astonishing experiments with young chicks, the French researcher René Peoc'h has demonstrated just such an effect. His experiments involved young chicks bonding to a machine instead of their mother.

Newly hatched chicks, like newly hatched ducklings and goslings, "imprint" on, or form an attachment to, the first moving object they encounter, and they then follow it around. Under normal circumstances, this imprinting instinct causes them to bond with their mother, but if the eggs are hatched in an incubator and young birds first meet a person, they will follow that person around instead. In laboratory experiments they can even be induced to imprint on moving balloons or other inanimate objects.

In his experiments, Peoc'h used a small robot that moved around on wheels in a series of random directions. At the end of each movement, it stopped, rotated through a randomly selected angle, and moved in a straight line for a randomly determined period before stopping and rotating again, and so on. These movements were determined by a random-number generator inside the robot. The path it traced out was recorded. In control experiments, its movements were indeed haphazard.

Peoc'h exposed newly hatched chicks to this robot, and they imprinted on this machine as if it were their mother. Consequently they wanted to follow it around, but Peoc'h stopped them from doing so by putting them in a cage. From the cage the chicks could see the robot, but they could not move toward it. Instead, they made the robot move toward them (Figure 16.1). Their desire to be near the robot somehow influenced the random-number generator so that the robot stayed close to the cage.[5] Chicks that were not imprinted on the robot had no such effect on its movement.

In other experiments, Peoc'h kept non-imprinted chicks in the dark. He put a lit candle on the top of the robot and put the chicks in the cage where they could see it. Chicks prefer being in the light during the daytime, and they "pulled" the robot toward them, so that they received more light.[6]

Peoc'h also carried out experiments in which rabbits were put in a cage where they could see the robot. At first they were frightened of it, and the robot moved away from them; they repelled it. But rabbits exposed to the robot daily for several weeks were no longer afraid of it and tended to pull it toward them.[7]

Thus the desire or fear of these animals influenced random events at a distance so as to attract or repel the robot. This would obviously not be possible if animals' desires and fears were confined inside their brains. Instead, their intentions reached out to affect the behavior of this machine.

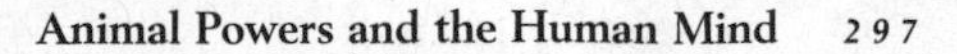

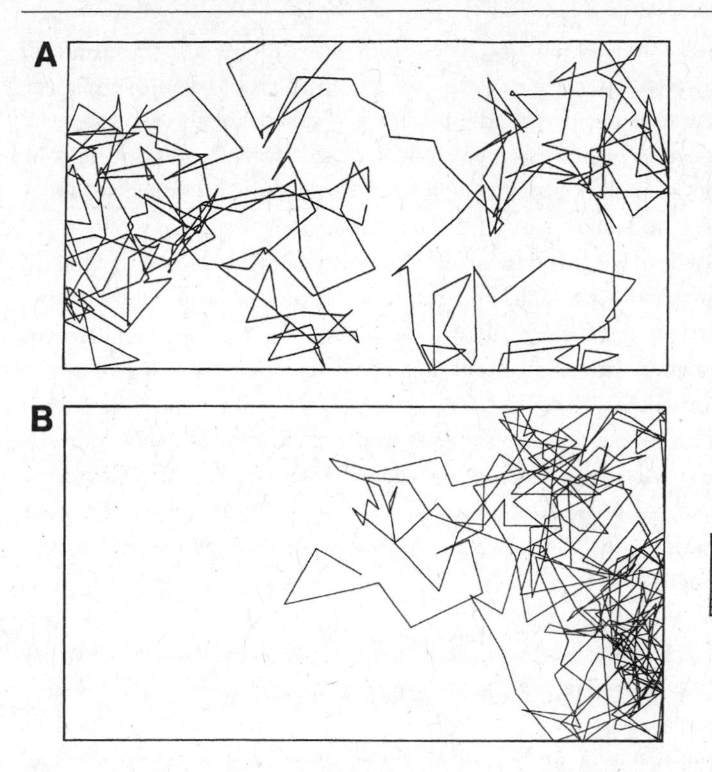


***Figure 16.1** The path traced out by the moving robot in experiments of René Peoc'h. A: A control experiment in which the cage was empty. B: An experiment in which day-old chicks imprinted on the robot were kept in the cage. (Reproduction courtesy of René Peoc'h)*

I interpret this influence in terms of a morphic field that projects out to the focus of their attention, connecting them to it. Just as a field of intention can affect people or animals at a distance, so it can affect a physical system. In one case, intention has effects at a distance on the brain. In the other case, intention has effects on random events in a machine.

As far as I know, no one has yet repeated Peoc'h's experiments. It is possible that they involve some technical flaw that no one has yet spotted. But if they are reliable and repeatable, they are very important indeed.

There is already good evidence from experiments at Princeton and other universities that people can indeed bring about mind-over-matter effects at a distance on random-event generators linked to computers. These devices produce the equivalent of electronic heads and tails in a random sequence, like tossing a coin. Participants are asked to try to influence the system so that, in a given period, there are more heads than tails or more tails than heads. These experiments have given highly significant and repeatable positive results, although the effects themselves are small. People really can influence chance events at a distance in accordance with their intention, although some people are better at it than others.[8]

Peoc'h's pioneering experiments imply that animals, both domesticated and wild, may well affect what happens around them through their fears and desires. But no one knows how great the power of animal intention may be. Nor do we know how great is the power of our own intentions.[9]

## The feeling of being stared at

Intentions reaching out beyond the brain may also give rise to a feeling of being looked at. Most people have, on occasion, felt other people looking at them from behind, and most have also looked at others from behind and found that they turned around. Surveys show that between 75 percent and 97 percent of Americans and Europeans say they have experienced the feeling of being looked at from behind.[10]

All around the world, there are folktales about the power of a look. In India people will travel hundreds of miles for the blessing conferred by the look, or *darshan*, of a holy man or woman. Conversely, people in many countries believe that a look of anger or envy, called the evil eye, can blight whatever it falls upon. Throughout the world, people take protective measures against the evil eye through prayers, charms, talismans, and amulets.[11] The idea that a malevolent glance can do grievous harm to person and property is so ancient that it is mentioned in the Bible as well as in Sumerian and other Near Eastern texts.[12]

Precisely because such beliefs are so common, most scientists treat them as superstitions, unworthy of serious consideration. They are denied or dismissed. Nevertheless, most people take these experiences for granted, and the sense of being stared at is well known to people who watch others professionally, as I have found through an extensive series of interviews with detectives, surveillance personnel, and soldiers. Most were convinced of the reality of this sense and told stories about times when people they were watching seemed to know they were being observed, however well the observers were hidden.[13] When detectives are trained to follow people, they are told not to stare at their backs any more than necessary, because otherwise the person might turn around, catch their eye, and blow their cover. In some of the Asian martial arts, students are trained to increase their sensitivity to being looked at from behind.

Many species of nonhuman animals also seem able to detect looks. Some pet owners claim that they can wake their sleeping dogs or cats by staring at them. Some hunters and wildlife photographers are convinced that animals can detect their gaze even when they are hidden and looking at the animals through telescopic lenses or sights. Conversely, some photographers and hunters say they have felt when they were being looked at by wild animals.[14] The naturalist William Long described how, when sitting in the woods alone as a boy, "I often found within myself an impression, which I expressed in the words, 'Something is watching you.' Again and again, when nothing stirred in my sight, that curious warning would come; and almost invariably, on looking around, I would find some bird or fox or squirrel that had probably caught a slight motion of my head and had halted his roaming to creep near and watch me inquisitively."[15]

In a survey in Ohio, Gerald Winer and his colleagues at Ohio State University found that many people said they have sensed the looks of animals. In that survey, 34 percent of adults and 41 percent of children said that they had felt when animals were looking at them. About half the respondents believed that animals could feel their looks, even when the animals could not see their eyes.[16]

If the sense of being stared at is so widespread, then it must have been subject to evolution by natural selection. The most obvious possibility is that it evolved in the context of predator-prey relations. Prey animals that could detect when predators were looking at them would probably stand a better chance of surviving than those that could not.[17]

In spite of the widespread familiarity of this sense, until the late 1980s there was very little research on this subject, even by parapsychologists. I have been able to find only six reports of experimental investigations between 1885 and 1985, including two in unpublished student theses. The four most recent gave positive, statistically significant results. The main reason for the persistent neglect of this phenomenon has nothing to do with evidence or experience. It flows from a belief that the sense of being stared at is impossible. Materialists believe that the mind is nothing but the activity of the brain; all mental activity is confined to the inside of the head, and thoughts and intentions should have no mysterious effects at a distance. Therefore the sense of being stared at cannot exist and must be denied. This is not a scientific hypothesis but an ideological belief.

Fortunately the feeling of being stared at can be investigated by means of simple, inexpensive experiments.[18] In these experiments, people work in pairs, with one person wearing a blindfold and sitting with his or her back to the other. The other person either looks at the back of the subject's neck or looks away. In a series of trials the sequence of looking and not-looking periods is randomized. In each trial the person wearing the blindfold has to guess whether he or she is being looked at or not. The guess is either right or wrong, and the scores are recorded. To date, tens of thousands of trials have been carried out with results that are overwhelmingly positive and highly significant statistically.[19] This effect still occurs when people are looked at through one-way mirrors. Far from being a superstition, the ability to detect stares seems to be real.

Surprisingly, people even seem to be able to detect stress when they are watched through closed-circuit television (CCTV). Millions of CCTV cameras are routinely used for surveillance in shopping malls, banks, offices, airports, streets, and other public spaces. My assistants and I have interviewed a representative sample of surveillance officers and security personnel whose job is to observe people through CCTV systems. Most, but not all, were convinced that some people could tell when they were being watched, and they gave examples to support this opinion.[20] However, to be taken seriously, anecdotal evidence for the sense of being stared at through CCTV would need to be supported by evidence from controlled experiments.

Such experiments have been performed with subjects and lookers in separate rooms. In these tests the subjects were not asked to guess whether they were being looked at or not. Instead, their galvanic skin response was recorded automatically, as in lie-detector tests. In a randomized series of trials the lookers either looked at the subjects' image on the TV monitor or looked away and thought of something else. Most of these experiments gave statistically significant positive results. The subjects' skin resistance changed when they were being looked at, even though they were unconscious of this change.[21]

These experiments show that the mind seems capable of reaching out to influence what is at its focus of attention. Indeed, the very word *attention* implies such a process. Its Latin roots have the meaning of stretching the mind toward something: *ad* means "toward," and *tendere* means "to stretch." It is closely related to the world *intention,* which means "to stretch the mind *into* something."

## Explaining the unexplained powers of animals

Throughout this book, I have discussed a variety of unexplained animal powers, and I have suggested that the idea of morphic fields could help explain many of them. But it cannot explain them all.[22]

For me, the most mysterious kinds of perceptiveness are those premonitions that cannot be accounted for in terms of telepathy or subtle physical clues. In these cases, by a process of elimination, precognition or presentiment seem the only remaining possibility. In precognition and presentiment, events about to happen somehow seem to influence animals now, alerting them to potential danger.

I do not pretend to know how animals' knowledge of the future might work. But at the very least, the existence of precognition or presentiment implies a blurring of what is happening now and what is about to happen.

There is a continuity between past, present, and future, as we know from our own experience, and as science assumes as a basis for understanding the course of nature. But the conventional scientific assumption is that influences work only from the past. Cause precedes effect. Energy and causation flow from past to present, and from present to future. There is supposed to be no flow of influence in the reverse direction.

The existence of precognition would imply that the conventional assumption is wrong, with huge implications for our understanding of mind, time, and causation.[23]

One way of thinking about precognition is to suppose that there are time-reversed information flows. An alternative is to reexamine our normal concept of the present. Perhaps this is too limited. What we call "now" is a moment that has a certain "thickness" in the space-time continuum, some fraction of a second. But what we consciously experience as "now" may be much shorter than what unconscious parts of ourselves experience as "now." In presentiment experiments, people became physiologically aroused a few seconds *before* they saw an emotionally stimulating image, suggesting that the present may indeed be thicker than our conscious awareness.

There are big issues at stake, and there is much we do not know. In order to understand more about the way forebodings and precognitions might work, I believe we need to start from a better-documented natural history of animal and human premonitions.

## Invisible interconnections

The development of science has involved the progressive recognition of invisible interconnections between things that are separated from each other in space or space-time. The concept of morphic fields takes this process further.

Modern science began in the seventeenth century with a grand vision of universal interrelationship. According to Isaac Newton's theory of gravitation, the earth attracts the moon through empty space, and the moon attracts the earth, as revealed by its influence on the tides. Likewise, the sun attracts the earth, and the earth attracts the sun. Indeed every material body in the universe attracts every other body: everything is interconnected.

Then there are the magnetic fields of the earth, the sun, and all other magnetic bodies, stretching out far beyond the material bodies themselves. Look at a compass when you are flying in a plane at 30,000 feet and it will still point north. The magnetic field of the earth pervades the space around it.

Radiation travels to the earth from distant galaxies through electromagnetic fields extending over billions of light-years. On earth, electromagnetic fields can connect us invisibly to events in distant places, as our experience of radio, television, and mobile telephones continually reminds us. The room you are in is full of radiations from thousands of radio and television transmitters. You are surrounded by vast amounts of invisible information, whether you have a receiver that can tune into it or not.

All these kinds of interconnection are now beyond dispute. They are the basis of the modern technologies on which we all depend. We take them for granted. It is easy to forget that they would have been inconceivable only generations ago. Who in the eighteenth century could have imagined television or the World Wide Web? But physics has already gone further.

According to quantum theory, there is an inevitable link between the observer and that which is observed, breaking down the sharp separation between subject and object. Scientists are no longer detached observers, seeing reality, as it were, through a plate-glass window. They participate in the reality they are studying. As one physicist has written, "We can no longer maintain the old Cartesian view that we can observe nature like a bird-watcher with a perfect hide. There is an unbreakable connection between the observer and the observed."[24]

Even more surprisingly, according to quantum physics, particles that come from a common source, like two photons of light emitted from the same atom, retain a mysterious interconnection so that what happens to one is instantaneously reflected in the other. This is known as non-locality, non-separability, or entanglement. No one knows how far this process extends or how extensive is this instantaneous interconnectedness. Some physicists speculate that everything in the universe is interconnected through quantum non-locality. According to Paul Davies and John Gribbin, "Once two particles have interacted with one another they remain linked in some way, effectively parts of the same indivisible system. This property of 'non-locality' has sweeping implications.

We can think of the universe as a vast network of interacting particles, and each linkage binds the participating particles into a single quantum system."[25]

## Morphic fields

Morphic fields also connect parts of a system that are seemingly separated, although no one yet knows how they are related to quantum non-locality. These fields are the basis of interconnections not only in space but also in time.

A wide range of unexplained powers of animals might be explicable in terms of morphic fields:

- Morphic fields link members of social groups and can continue to connect them even when they are far apart (Figure 1.5, page 25). These invisible bonds act as channels for telepathic communication between animals and animals, people and animals, and people and people.
- These links, acting like invisible elastic bands, also underlie the sense of direction that enables animals and people to find each other (Chapter 13).
- Animals "imprinted" on their home environment or on other significant places are linked to these places by morphic fields. Through these connections they can be pulled or attracted back toward familiar places, enabling them to navigate across unfamiliar terrain. The sense of direction given by these morphic fields underlies both homing and migration.
- Morphic fields link animals to the objects of their intentions, and could help to explain psychokinetic phenomena.
- Morphic fields link animals to the objects of their attention, and through these perceptual fields animals can influence what they are looking at. These fields underlie the sense of being stared at.

Thus the idea of morphic fields may be capable of giving a unified explanation to a wide range of seemingly disparate phenomena.

Other people may prefer to call these fields by different names or to use words like "system" or "interrelationship" instead of "field." But whatever such interconnections are called, I expect that they will have most of the properties of morphic fields.[26]

## Learning from our animals

Whatever explanations turn out to be the best, there is no doubt that we have much to learn from our dogs, cats, horses, parrots, pigeons, and other domesticated animals. They have much to teach us about social bonds and animal perceptiveness, and much to teach us about ourselves.

The evidence I have discussed in this book suggests that our own intentions, desires, and fears are not confined to our heads or communicated only through words and behavior. We can influence animals and affect other people at a distance. We remain interconnected with animals and people we are close to, even when we are far away. We can even affect people and animals by the way we look at them, even if they do not know we are there. We can retain a connection with our homes, however geographically distant we are. And we can be influenced by things that are about to happen.[27]

These are areas of investigation that most scientists have ignored for generations. There is a taboo against them because they do not fit in to the mechanistic theory of nature, on which science has been based for four hundred years. Some people are afraid that taking these phenomena seriously will encourage superstition and undermine the hard-won advances of civilization. This fear motivates many members of skeptical organizations, whose mission is to debunk the paranormal in the name of science and reason.

I believe that science should be a method of open-minded inquiry rather than a dogmatic belief system. I am convinced that telepathy, the sense of direction, and premonitions are normal, not paranormal. They invite us to widen our view of nature and expand the scope of science and reason.

*Appendix*

# Controversies and Inquiries

The topics discussed in this book are considered taboo by some members of the scientific world. The very ideas of telepathy, an unexplained sense of direction, premonitions, or precognitions arouse skepticism, if not hostility, among some scientists and philosophers. Why?

My research has led me into a series of intense controversies. People with no experience of professional science may imagine that it is all about the open-minded exploration of the unknown, but this is rarely the case. Science works within frameworks of belief or models of reality. Whatever does not fit in is denied or ignored; it is anomalous. The historian of science Thomas Kuhn called these thought-patterns paradigms. During periods of what he called normal science, scientists work within the paradigm and ignore or deny anomalies.

In scientific revolutions orthodox paradigms are challenged and replaced with new, larger models of reality that can incorporate previously rejected anomalies. In due course these new thought patterns become standard orthodoxies.[1]

The paradigm that has dominated institutional science since the nineteenth century is materialism: Matter is the only reality. Mind or consciousness exists only insofar as it arises from material processes in brains. Animals—and people—are nothing but complex machines, explicable in terms of the ordinary laws of physics and chemistry. Minds are inside brains and cannot have mysterious effects at a distance.

Ironically, although materialists put their faith in physical laws, these laws are not themselves physical. They are conceived of as nonmaterial principles that transcend space and time, potentially active at all times and in all places. Moreover, several modern physicists have pointed out that nothing in modern physics—as opposed to nineteenth-century physics—would be compromised by the existence of abilities such as telepathy. In the light of quantum theory, the laws of classical physics have been rewritten.[2]

The topics explored in this book are anomalies from the point of view of materialism. That is why they are controversial. If they really exist, then they point toward a new, larger model of reality, a new paradigm.

For most believers in materialism, God is nothing but a delusion inside human minds, and hence inside heads. People with a strong materialist faith are usually atheists as well. Atheists are not people without a belief; they are people with a strong faith in the doctrine of materialism. From their point of view religious beliefs are nonsensical and so are psychic phenomena. During what is somewhat arrogantly called the Enlightenment, the materialism and determinism of classical science gave intellectuals the tools to challenge the authority of church and scripture with the authority of science. Modern secular humanists are the direct descendants of the Enlightenment thinkers, and their worldview is for the most part still based on the materialism implied by classical physics. If materialism is falsified by the data for telepathy and other psychic phenomena, then one of the foundations of their opposition to religion is thereby removed. Hence the vehement denial of any evidence for the existence of phenomena that go against their beliefs.[3]

Regardless of what materialists think, most people believe that they have had telepathic experiences, often in connection with telephone calls, by thinking of someone who then calls. Many owners of dogs, cats, horses, parrots, and other animals find their animals pick up their thoughts and intentions. Some scientists have telepathic experiences themselves, and some have dogs that know when they are coming home from the laboratory. But scientists usually keep quiet about these experiences. At work they function within a materialist paradigm; in their private lives many are religious, or follow spiritual paths, or think there is more in heaven and earth than is dreamed of in the materialist philosophy. Only a minority are card-carrying atheists.

Not all atheists are opposed to research on psychic phenomena. Sam Harris, author of *The End of Faith*, is open to the possibility that some of these phenomena may be real. Meanwhile several eminent parapsychologists are atheists. They hope that psychic phenomena can be incorporated into an enlarged scientific model of reality. I share that hope, although I am not an atheist myself.

Unfortunately, much passion arises because materialists feel that science and reason themselves are being threatened by acknowledging the existence of psychic phenomena. But that is only the case if science is identified with old-style materialism. There is an alternative scientific possibility: Psychic phenomena are compatible with an expanded scientific model of reality and are independent of the question of the existence of God. Psychic phenomena like telepathy are natural, not supernatural. They no more prove or disprove the existence of God than do the sense of smell or the existence of electromagnetic fields.

## Skepticism

Genuine skepticism is healthy and an integral part of science. Scientists in all areas of professional research are subjected to institutionalized skepticism in the form of anonymous peer review. Whenever they submit a paper to a scientific journal, the editor sends it to two or more referees, often the authors' competitors or rivals, whose names are not revealed to the author. This is normal scientific practice, and I am used to it after publishing more than eighty papers in peer-reviewed journals. Grant proposals are often peer-reviewed as well.

However, another kind of skepticism comes into play in relation to taboo topics like telepathy—the dogmatic skepticism of people defending a belief system or orthodoxy. The more militant the skeptic, the more passionate the belief.

In the United States, Britain, and many other countries there are a variety of skeptic organizations that see it as their job to debunk what they call claims of the paranormal. In the United States, the largest of these groups is called the Committee for Skeptical Inquiry (CSI), which until 2006 was called the Committee for the Scientific Investigation of Claims of the Paranormal (CSICOP). Its magazine, *Skeptical Inquirer*, has approximately 50,000 subscribers. CSICOP was founded in 1978 at a meeting of the American Humanist Association by Paul Kurtz, an atheist philosopher, who also founded the Council for Secular Humanism and the Center for Inquiry. CSICOP/CSI shares its headquarters in Amherst, New York, with the Council for Secular Humanism as well as with an organization devoted to debunking alternative medicine, the Commission for Scientific Medicine and Mental Health (CSMMH). CSI has about eighty Fellows, including militant atheists such as Richard Dawkins, author of *The God Delusion*, and Daniel Dennett, author of *Breaking the Spell*.

When Paul Kurtz announced the change of name from CSICOP to CSI in the January 2007 *Skeptical Inquirer*, he looked back over CSICOP's past and made it clear that the organization's agenda was rooted in an ideological commitment: "We viewed ourselves as the defenders of the Enlightenment." In an interview for *Science* magazine, Lee Nisbet, the CSICOP Executive Director, put it as follows: "[Belief in the paranormal is] a very dangerous phenomenon. Dangerous to science, dangerous to the basic fabric of our society. . . . We feel it is the duty of the scientific community to show that these beliefs are utterly screwball." As is the case with so many of the leading figures in CSICOP/CSI, Nisbet has no scientific qualifications.

CSICOP/CSI's primary efforts are directed to influencing public opinion. The *Skeptical Inquirer* carries innumerable articles decrying the media's treatment of the paranormal and describes CSICOP's attempts to combat any favorable coverage. As reported in the *Skeptical Inquirer*,[4] CSICOP originated "to fight mass-media exploitation of supposedly 'occult' and 'paranormal' phenomena. The strategy was twofold: First, to strengthen the hand of skeptics in the media by providing information that 'debunked' paranormal wonders. Second, to serve as a 'media-watchdog' group that would direct public and media attention to egregious media exploitation of the supposed paranormal wonders. An underlying principle of action was to use the main-line media's thirst for public-attracting controversies to keep our activities in the media, hence in the public eye. Who thought this strategy up? Well, Paul Kurtz, that's who."

In a penetrating essay called "The Skepticism of Believers," published in 1893, Sir Leslie Stephen, a pioneering agnostic (and the father of the novelist Virginia Woolf), argued that skepticism is inevitably partial. "In regard to the great bulk of ordinary beliefs, the so-called skeptics are just as much believers as their opponents." Then as now, those who proclaim themselves skeptics had strong beliefs of their own. As Stephen put it, "The thinkers generally charged with skepticism are equally charged with an excessive belief in the constancy and certainty of the so-called laws of nature. They assign a natural cause to certain phenomena as confidently as their opponents assign a supernatural cause."

Almost all the people who have attacked me as a result of the research with animals described in this book have been Fellows of CSICOP, militant atheists, or career skeptics, not professionals who actually know about animals: researchers in animal behavior, animal trainers, or vets. I have given seminars in veterinary schools and lectured at academic conferences on companion animals (the academic term for pets) to audiences who seemed genuinely interested in the studies described in this book. I have spoken on this research in dozens of university science and psychology departments; to student science societies; at international scientific conferences; to scientific institutes, including the European Molecular Biology Laboratory, in Heidelberg, Germany; at international conferences on consciousness studies; and to corporations like Microsoft, Nokia, and Google. (My technical seminar at Google is online on the Google website.[5]) Of course some of the people at these events have been skeptical, but again and again I have found that dogmatic skeptics are a small minority. They often claim to speak for the scientific community, but fortunately most scientists are more open-minded.

Here is a summary of some of my encounters.

## Sir John Maddox, editor of *Nature*

The late Sir John Maddox, one of CSICOP's most eminent Fellows, was my longest-standing critic. As the editor of *Nature*, the prestigious scientific journal, he was the author of an infamous *Nature* editorial about my first book, *A New Science of Life*, in which he wrote, "This infuriating tract . . . is the best candidate for burning there has been for many years." In an interview broadcast on BBC television in 1994 he said, "Sheldrake is putting forward magic instead of science, and that can be condemned in exactly the language that the Pope used to condemn Galileo, and for the same reason. It is heresy."[6]

Maddox reviewed *Dogs That Know When Their Owners Are Coming Home* in *Nature* in October 1999.[7] This is how he began: "Rupert Sheldrake is steadfastly incorrigible in the particular sense that he persists in error. That is the chief import of his eighth and latest book. Its main message is that animals, especially dogs, use telepathy in routine communications. The interest of this case is that the author was a regular scientist, with a Cambridge Ph.D. in biochemistry, until he chose pursuits that stand in relation to science as does alternative medicine to medicine proper."

Maddox alluded to his attack on my first book, paraphrased my ideas about morphic fields and morphic resonance, and traced their development over the years. He gave an overview of *Dogs That Know When Their Owners Are Coming Home,* and summarized some of the experiments with Jaytee. He then raised a number of questions:

> By conceding that the data gathered during these observations are statistically significant, one does not sign up for Sheldrake's interpretation that the underlying mechanism is dog-*Homo* telepathy. Too many variables are uncontrolled. Did the accuracy of anticipation vary with the length of time elapsed since Pam's departure (suggesting that the dog used its sense of the passage of time to signal its sense of when return was due)? Were there people in the room with the dog (allowing them to communicate somehow with the eager waiter)? And while Jaytee appears to have been chosen for videotaping as a result of his acumen in earlier trials, does not the interpretation of his behavior require an understanding of the variability of dogs' capacity for anticipation in general?

Maddox concluded his review as follows:

> Especially because people's fondness for their pets often takes the form of projecting onto them human or even superhuman perceptiveness, even more than 1,000 records on the Sheldrake website do not prove telepathy. I doubt that Sheldrake will take the point. He makes plain his distaste for what he calls orthodox science, which is

> "all too often equated with a narrow-minded dogmatism that seeks to deny or debunk whatever does not fit in with the mechanistic view of the world." He is habitually courteous and cheerful, but holists of his ilk would not dream of letting controls get in the way of revealed truth.

I wrote to Maddox taking up the scientific points he raised, starting with his suggestion that Jaytee used the passage of time to signal when Pam was returning. I pointed out, "The longer the absence, the longer the time the dog took to start waiting at the door when Pam was on her way home. A statistical analysis comparing the long, medium, and short experiments ruled out the passage of time argument. So did the control experiments carried out when Pam was not coming home. So I think this question is already answered by the data."

Maddox's second question was about people in the room with the dog. I wrote, "As I make clear in my account, in experiments at Pam's parents' flat, her parents were in the room, but since they did not know when she was coming home, especially in the experiments with randomized return times, the only way they could have communicated this information to her would be if they themselves picked up telepathically when she was on her way. In experiments at Pam's sister's house, her sister was present, but again only a person-to-person telepathy argument would provide a real alternative explanation. And then we carried out fifty experiments with the dog alone in Pam's flat. He still showed his reactions to a statistically significant extent when completely alone."

The third question about the variability for dogs' capacity for anticipation in general was obscure, or at least too vague to answer, though I had much data on dogs' anticipatory behavior in general. I ended my letter to Maddox as follows:

> In your final remark you say, "Holists of his ilk would not dream of letting controls get in the way of revealed truth." If you mean other unspecified persons, then it is meaningless and irrelevant. If you mean me, then what you say is unjust and untrue. I have done thousands of experiments over the years involving controls, as you can see by looking at my many published papers. And of course I use controls in my research with animals. I have never regarded animal telepathy

as revealed truth; it is certainly no article of faith for any religion, nor is it even mentioned in most books on parapsychology. I entered this field of inquiry with an open mind about what animals can and cannot do, and would not otherwise have spent years in empirical investigations of their abilities.

Maddox did not reply, although when I met him several months later at a seminar at the Royal Society, he said, "I ought to have replied to your letter but I haven't got 'round to it." He died in 2009 and never got 'round to it.

## James "The Amazing" Randi

James Randi is a showman, conjurer, and a former Principal Investigator of CSICOP. For years he frequently appeared in the media as a debunker of the paranormal. He was named "Skeptic of the Century" in the January 2000 issue of the *Skeptical Inquirer,* and in 2003 received the Richard Dawkins Award from Atheist Alliance International.

In 1996 he founded the James Randi Educational Foundation (JREF) and is most famous for offering a million-dollar paranormal challenge to anyone who can demonstrate evidence of a paranormal event under conditions to which he agrees. Randi has no scientific credentials and has disarmingly said of himself, "I'm a trickster, I'm a cheat, I'm a charlatan, that's what I do for a living."[8]

In January 2000 *Dog World* magazine published an article on the sixth sense of dogs, which discussed my research. The author contacted Randi to ask his opinion. Randi was quoted as saying that in relation to canine ESP, "We at the JREF have tested these claims. They fail." Randi also claimed to have debunked one of my experiments with Jaytee, in which Jaytee went to the window to wait for his owner when she set off to come home at a randomly selected time but did not go to the window before his owner left to come home. In *Dog World* Randi stated, "Viewing the entire tape, we see that the dog responded to every car that drove by and to every person who walked by."

I e-mailed James Randi to ask for details of this JREF research. He did not reply. He ignored a second request for information. I then asked members of the JREF Scientific Advisory Board to help me find out more about this claim. They advised Randi to reply.

In an e-mail on February 6, 2000, Randi told me that the tests with dogs he referred to were not done at the JREF but took place "years ago" and were "informal." He said they involved two dogs belonging to a friend of his that he observed over a two-week period. All records had been lost. He wrote: "I overstated my case for doubting the reality of dog ESP based on the small amount of data I obtained."

I also asked him for details of the tape he claimed to have watched, so I could compare his observations of Jaytee's behavior with my own. He was unable to give a single detail, and under pressure from the JREF Advisory Board he had to admit that he had never seen the tape. His claim was a lie.

For many years the million-dollar prize has been Randi's stock-in-trade as a media skeptic, but even other skeptics are skeptical about its value as anything but a publicity stunt. For example, CSICOP founding member Dennis Rawlins pointed out that Randi acts as "policeman, judge, and jury," and he quoted him as saying, "I always have an out."[9] Ray Hyman, a professor of psychology and Fellow of CSICOP, pointed out this "prize" cannot be taken seriously from a scientific point of view: "Scientists don't settle issues with a single test, so even if someone does win a big cash prize in a demonstration, this isn't going to convince anyone. Proof in science happens through replication, not through single experiments."[10]

Nevertheless I asked the Smart family if they would be willing to have Jaytee tested by Randi. But they wanted nothing to do with him. Jaytee had already taken part in some tests organized by a skeptic, Richard Wiseman, as discussed below, and the Smart family was disgusted by the way he had misrepresented these tests in the media.

In 2008 Alex Tsakiris, who runs a U.S.-based Open Source Science Project and a podcast called Skeptiko, started replicating experiments with dogs that knew when their owners were coming home, posting videos of tests on the Internet. Tsakiris asked Dr. Clive Wynne, an expert on dog behavior at the University of Florida, to participate in this research, and Wynne agreed. Randi challenged Tsakiris to apply for the million-dollar challenge; Tsakiris took him up on it and asked Randi by e-mail if Dr. Wynne's involvement was acceptable to him. Randi eventually replied, "You appear to think that your needs are uppermost on my schedule. What would give you that impression? Looking into a silly dog claim is among my lowest priority projects. When I'm prepared to give you some time, I'll let you know. There are some forty-plus persons ahead of you."[11]

For me the most surprising feature of the Randi phenomenon is that so many journalists and fellow skeptics take him seriously.

## Richard Wiseman

Richard Wiseman started his career as a conjurer and like Randi is a skilled illusionist. He has a Ph.D. in psychology and is an expert on the psychology of deception. He is a Fellow of CSICOP/CSI, one of Britain's best-known media skeptics, and is currently Professor of the Public Understanding of Psychology at the University of Hertfordshire.

When my experiments with Jaytee were first publicized in Britain in 1994, journalists sought a skeptic to comment on them, and Richard Wiseman was an obvious choice. He put forward a number of points that I had already taken into account, suggesting that Jaytee was responding to routines, or car sounds, or subtle cues. But rather than argue academically, I suggested that he carry out some experiments with Jaytee himself, and I arranged for him to do so. I had already been doing videotaped experiments with this dog for months, and I lent him my video camera. Pam Smart, Jaytee's owner, and her family kindly agreed to help him. Along with his assistant, Matthew Smith, he did four experiments with Jaytee, two in June and two in December 1995, and in all of them Jaytee went to the window to wait for Pam when she was indeed on the way home.

As in my own experiments, Jaytee sometimes went to the window at other times—for example, to bark at passing cars—but he was at the window far more when Pam was on her way home than when she was not. In the three experiments Wiseman did in Pam's parents' flat, Jaytee was at the window an average of 4 percent of the time during the main period of Pam's absence and 78 percent of the time when she was on the way home. This difference was statistically significant. When Wiseman's data were plotted on graphs they showed essentially the same pattern as my own (Figure 2.5). In other words Wiseman replicated my own results.

I was astonished to hear that in the summer of 1996 Wiseman went to a series of conferences, including the World Skeptics Congress, announcing that he had refuted the psychic pet phenomenon. He said Jaytee had failed his tests because he had gone to the window before Pam set off to come home. In September 1996 I met Wiseman and pointed out that his data showed the same pattern as my own, and that far from refuting the effect I had observed his results confirmed it. I gave him copies of graphs showing my own data and the data from the experiments that he and Smith conducted with Jaytee. But he ignored these facts.

Wiseman reiterated his negative conclusions in a paper in the *British Journal of Psychology,* coauthored with Smith and Julie Milton, in August 1998.[12] This paper was announced in a press release entitled "Mystic dog fails to give scientists a lead," together with a quote from Wiseman: "A lot of people think their pet might have psychic abilities, but when we put it to the test what's going on is normal not paranormal." There was an avalanche of skeptical publicity, including newspaper reports with headlines like "Pets have no sixth sense, say scientists" (*The Independent,* August 21) and "Psychic pets are exposed as a myth" (*The Daily Telegraph,* August 22). Smith was quoted as saying, "We tried the best we could to capture this ability and we didn't find any evidence to support it." The wire services reported the story worldwide. Skepticism appeared to have triumphed.

Wiseman continued to appear on TV shows and in public lectures claiming he had refuted Jaytee's abilities. Unfortunately, his presentations were deliberately misleading. He made no mention of the fact that in his own tests Jaytee waited by the window far more when Pam was on her way home than when she was not, nor did he refer to my own experiments. He gave the impression that my evidence was based on one experiment filmed by a TV company, rather than on more than two hundred tests, and he implied that he has done the only rigorous scientific tests of this dog's abilities.

Instead of plotting their data on graphs and looking at the overall pattern, Wiseman, Smith, and Milton used a criterion of their own invention to judge Jaytee's success or failure. They did not discuss this criterion with me, although I had been studying Jaytee's behavior in detail for more than a year before I invited them to do their own tests. They instead based their findings on remarks about Jaytee's behavior made by commentators on two British television programs, who said that Jaytee went to the window every time that his owner was coming home, when in fact he did so on 86 percent of the occasions.[13] And one of these programs said that Jaytee went to the window "when his owner, Pam Smart, starts her journey home." In fact Jaytee often went to the window a few minutes *before* Pam started her journey, while she was preparing to set off.[14] Based on these TV commentaries, Wiseman, Smith, and Milton took Jaytee's "signal" to be the dog's first visit to the window for no apparent external reason. They later changed this criterion to a visit that lasted more than two minutes.

Wiseman and Smith found that Jaytee sometimes went to the window at Pam's parents' flat for no obvious reason before Pam set off at the randomly selected time. Anytime this happened they classified the test as a failure, despite the fact that Jaytee waited at the window 78 percent of the time when Pam was on the way home, compared with only 4 percent when she was not. They simply ignored the dog's behavior after the "signal" had been given. In addition to these experiments at Pam's parents' flat, they carried out a test at the house of Pam's sister, where Jaytee had to balance on the back of a sofa to look out of the window. The first time he visited the window for no apparent reason coincided exactly with Pam setting off, and her sister remarked at the time, on camera, that this was how Jaytee behaved when Pam was coming home. But Jaytee did not stay there for long because he was sick; he left the window and vomited. Because he did not meet the two-minute criterion, this experiment was deemed a failure.

On another British television program called *Secrets of the Psychics,*[15] Wiseman said of Jaytee, "We filmed him continuously over a three-hour period, and at one point we had the owner randomly think about returning home from a remote location and yes, indeed, Jaytee was at the window at that point. What our videotape showed, though, was that Jaytee was visiting the window about once every ten minutes and so under those conditions it is not surprising he was there when his owner was thinking of returning home." To support this statement, a series of video clips showed Jaytee going to the window over and over again, eight times in all. The times of these visits to the window can be read from the time code. They were taken from the experiment on June 12 shown in Figure 2.5. Two of these visits were the same clip shown twice, and three took place while Pam was actually on the way home, although they were misleadingly portrayed as random events unrelated to her return. Looking at the graph of the data from this test, it is obvious that Jaytee spent by far the most time at the window when Pam was on the way home: He was there 82 percent of the time. In the previous periods his visits were much shorter, if he visited the window at all.

Wiseman, Smith, and Milton said that they were "appalled" by the way some of the newspaper reports portrayed Pam Smart.[16] But although they helped initiate this media coverage, they considered themselves blameless: "We are not responsible for the way in which the media reported our paper and believe that these issues are best raised with the journalists involved." They also excused themselves for failing to mention my own research with Jaytee on the grounds that it had not yet been published when they submitted their paper to the *British Journal of Psychology.* They therefore created the appearance that they were the only people to have done proper scientific experiments with a return-anticipating dog. Also by publishing their paper before I could publish my own—I spent two years doing experiments, while they spent four days—they claimed priority in the scientific literature for this kind of research. To put it mildly, these were scientific bad manners.

Wiseman still tells the media, "I've found plenty of evidence of unscientific approaches to data but have never come across a paranormal experiment that can be replicated."[17] In a comprehensive analysis of Wiseman's approach, Christopher Carter has shown how he adopts a "heads I win, tails you lose" approach to psychic phenomena, viewing null results as evidence against *psi* while attempting to ensure that positive results do not count as evidence for it. Carter has documented a series of examples, including the Jaytee case, where Wiseman uses "tricks to ensure he gets the results he wants to present."[18] He is, after all, an illusionist and an expert in the psychology of deception.

## Susan Blackmore

Dr. Susan Blackmore is a CSICOP/CSI Fellow, was awarded the CSICOP Distinguished Skeptic Award in 1991, and used to be one of Britain's best-known media skeptics. She started her career by doing research in parapsychology but left the field and later devoted herself to the study of memes, as proposed by Richard Dawkins.

Blackmore commented on my experiments with Jaytee in an article in the *Times Higher Education Supplement*,[19] claiming that she had spotted "design problems." She wrote, "Sheldrake did twelve experiments in which he beeped Pam at random times to tell her to return. . . . When Pam first leaves, Jaytee settles down and does not bother to go to the window. The longer she is away, the more often he goes to look. [Y]et the comparison is made with the early period when the dog rarely gets up." But anybody who looks at the actual data can see for themselves that this is not true.[20] In five out of the twelve experiments with random return times, Jaytee did not settle down immediately after Pam left. In fact he went to the window more in the first hour than during the rest of Pam's absence.

In the light of Blackmore's comments, I reanalyzed the data from all twelve experiments excluding the first hour. The percentage of time that Jaytee spent by the window in the main period of Pam's absence was actually lower when the first hour was excluded (3.1 percent) than when it was included (3.7 percent). By contrast, Jaytee was at the window 55 percent of the time when she was on the way home. Taking Blackmore's objection into account strengthened rather than weakened the evidence for Jaytee knowing when his owner was coming home and increased the statistical significance of the comparison.[21]

In addition, if Blackmore had taken the trouble to look at our data more thoroughly she would have seen that we did a series of control tests in which Pam did not come home at all. Jaytee did not go to the window more and more as time went on. See figure 5, page 247, in Sheldrake and Smart (2000).

Blackmore's claim illustrates once again the need to treat what skeptics say with skepticism.

## Michael Shermer

Michael Shermer is a professional skeptic rather than a scientist, although he often claims to speak for science. He is the publisher of *Skeptic* magazine, the Director of the Skeptic Society, the host of the *Skeptics'* Lecture Series at the California Institute of Technology, and the author of a regular column in *Scientific American* called "Skeptic." He started out as a Christian fundamentalist as well as being an enthusiast for pyramid power and other New Age fads. In his own words, "My academic background is embarrassing compared to that of most successful intellectuals. . . . I scraped together a master's degree . . . and finally gave up hope for an intellectual life and raced bikes for a decade. By the time I earned a Ph.D. [in history of science] . . . I discovered there were next to no jobs, especially for someone with an intellectual pedigree such as mine. Since teaching as an adjunct professor is no way to make a living (literally), I founded the Skeptics Society and *Skeptic* magazine."[22]

One of Shermer's favorite sayings is "Skepticism is a method, not a position." However, I soon discovered that he does not practice what he preaches. In 2003 *USA Today* published an article about my book *The Sense of Being Stared At*, describing my research on telepathy and the sense of being stared at. Shermer was asked for his comments and was quoted as saying, "[Sheldrake] has never met a goofy idea he didn't like. The events Sheldrake describes don't require a theory and are perfectly explicable by normal means."[23]

I e-mailed Shermer to ask him what his normal explanations were. But he was unable to substantiate his claim, and he admitted he had not even seen my book. I challenged him to an online debate. He accepted the challenge but said he was too busy to look at the experimental evidence and would "get to it soon." Several months later he confessed, "I have not gotten to your book yet." Despite repeated reminders, he has still failed to do so.

It only takes a few minutes to make an evidence-free claim to a journalist. Dogmatism is easy. It is harder work to consider the evidence, and Shermer is too busy to look at facts that go against his beliefs.

In November 2005 Shermer attacked me in his *Scientific American*

"Skeptic" column in a piece called "Rupert's Resonance."[24] He ridiculed the idea of morphic resonance by claiming that I proposed a "universal life force," a phrase I have never used. He also referred to fallacious, partisan claims by other skeptics about my experimental work, which had already been refuted in peer-reviewed journals and even in the *Skeptical Inquirer* itself.

I wrote a brief letter to *Scientific American* to set the record straight, but it was not published, nor even acknowledged, and Shermer himself ignored it. Other scientists whom Shermer has misrepresented have had the same experience.[25] The disciplines of science do not seem to apply to media skeptics.

The readers of *Scientific American* would be better served by a fair and truthful presentation of the facts than by Michael Shermer's misleading skepticism.

Meanwhile, Shermer continues to flatter himself with fine-sounding words. In 2010 he contrasted his kind of skepticism with denialism, as in climate change denial or holocaust denial or evolution denial: "When I call myself a skeptic, I mean I take a scientific approach to the evaluation of claims. . . . A climate denier has a position staked out in advance and sorts through the data employing 'confirmation bias'—the tendency to look for and find confirmatory evidence for preexisting beliefs and ignore or dismiss the rest. . . . Thus one practical way to distinguish between a skeptic and a denier is the extent to which they are willing to update their positions in response to new information. Skeptics change their minds. Deniers just keep on denying."[26]

By Shermer's own criteria he is a perfect example of a denier.

## Lewis Wolpert

Lewis Wolpert was Professor of Biology at University College, London, and served for five years as Chairman of COPUS, the British Committee on the Public Understanding of Science. He was a faithful standby for the media for more than twenty years as a denouncer of ideas that he suspected were tainted with mysticism or the paranormal.

In 2001 in a program about some of my telepathy experiments on the

Discovery Channel he proclaimed, "There is no evidence for any person, animal, or thing being telepathic."[27] The director of the documentary offered to show him a video of my experiments so that he could see the evidence for himself, but he was not interested. He preferred to make his skeptical claim without looking at the facts.

In January 2004 Wolpert and I took part in a public debate on telepathy at the Royal Society of Arts in London, with a high court judge in the chair. We were each given thirty minutes to present our cases. Wolpert spoke first and said that research on telepathy was "pathological science" and added, "An open mind is a very bad thing—everything falls out." He asserted that "the whole issue is about evidence" and concluded after a mere fifteen minutes that "there is zero evidence to support the idea that thoughts can be transmitted from a person to an animal, from an animal to a person, from a person to a person, or from an animal to an animal."

I then summarized evidence for telepathy from thousands of scientific tests and showed a video of recent experiments, but Wolpert averted his eyes from the screen. He did not want to know. According to a report on the debate in *Nature*, "few members of the audience seemed to be swayed by [Wolpert's] arguments. . . . Many in the audience . . . variously accused Wolpert of 'not knowing the evidence' and being 'unscientific.'"[28]

For anyone who wants to hear both sides for themselves, the debate is online in streaming audio, as is the transcript.[29]

## The European Skeptics Congress

I was invited to speak at the twelfth European Skeptics Congress in Brussels, Belgium, in October 2005. I took part in a plenary session in which there was a debate on telepathy between me and Jan Willem Nienhuys, the secretary of a Dutch skeptic organization, Stichting Skepsis. I presented evidence for telepathy, reviewing research by others and by myself. Nienhuys then responded by arguing that telepathy was impossible, and therefore all the evidence for it must be flawed. He commented that the more statistically significant my experimental results were, the greater the errors must be. I asked him to specify these errors, but he said he could not do so since he had not actually read my papers or studied the evidence.

Here is a description of the debate by an independent observer, Dr. Richard Hardwick, a scientist at the European Commission:

> Dr. Sheldrake was on first. . . . He came well prepared, and he spoke fluently and clearly, as if he really wanted to communicate. He marshaled his arguments with precision, he provided (so far as I can judge) evidence for his statements, and he brought his null hypotheses out into the open, ready to be shot down by the force of disproof. In my judgment, Nienhuys's counterattack failed. . . . It seems Dr. Nienhuys had not done his homework. He did not have any data or analyses at hand, and his attack fizzled out. So in the questionnaire that was (commendably) distributed to the participants for filling in afterward, I scored the encounter not "game, set, and match to Sheldrake," but at least "Sheldrake, 40; Nienhuys, love." A small cluster gathered around Sheldrake at the end of the Congress. They seemed to be talking with him, rather than pummeling him to the ground, so perhaps they agreed with me.[30]

## National Geographic TV Channel

The most flagrant example of a biased presentation of research with animals occurred on the National Geographic Channel in 2005. It was so bad that I complained about it to the British media regulator, the Government Office of Communications (Ofcom), whose duties include ensuring that television companies behave fairly. After considering my case, the response from National Geographic, and viewing the TV show, Ofcom issued an official adjudication ruling that National Geographic had broken the guarantee they had given me to present my research fairly. National Geographic was required to stop transmitting the program and to broadcast a summary of Ofcom's adjudication. National Geographic appealed against this decision, but the judge rejected all their arguments and upheld Ofcom's adjudication.[31] Meanwhile, in the United States, National Geographic Channel continued to repeat broadcasts of the offending program, called *Is It Real? Psychic Animals.*[32]

I was particularly disappointed by National Geographic's attitude since I had always held the National Geographic Society in high regard. My father subscribed to *National Geographic* magazine, which I read avidly throughout my childhood. I was a member of the society myself. But National Geographic is now a global brand, and the National Geographic Channel is largely owned by Fox Entertainment Group, part of Rupert Murdoch's media empire.

After the Ofcom adjudication, I wrote to the President of the National Geographic Society, John Fahey, asking him to stop further repeats of the program in the United States. As I expressed it, "This issue raises fundamental questions about honesty and integrity, and about the connection between the National Geographic Channel and the National Geographic Society, of which I am a member. As a respected educational institution, the interests of the National Geographic Society are not the same as those of its business partner, Fox Television: It has a reputation to protect. It was precisely because of this reputation that I agreed to take part in the first place and trusted the assurance given in the name of National Geographic." Fahey did not reply. Instead I received a letter from the legal department stating that Ofcom's findings would have no effect on their activities outside the United Kingdom.

Here is a summary of what happened. When I was asked to take part in the program, I was reluctant to do so because I was all too familiar with the debunking format that TV companies used when presenting controversial research. In the standard scenario, someone who had done serious research on unexplained phenomena was called a "proponent" making a "claim," and then a self-described skeptical investigator, usually with no scientific credentials, disdainfully debunked the claim. When I expressed my doubts the National Geographic producer, Dana Kemp, told me nothing about the involvement of the CSICOP team and replied as follows:

> We're used to the skeptical question—it's one that comes up a lot, and I understand the concern. Being National Geographic, and having a very strict policy of balanced reporting, we cannot be biased in either direction. It is our job to present the work being done, and where

deemed necessary and in all fairness we will often include the flip side of the coin. I will tell you that this is the first show I will be producing for this series, and as the producer I absolutely have no intention of putting anyone in an unfair, uncomfortable position or making anyone look silly. My goal is to present science.[33]

Contrary to these assurances the show was strongly biased toward dogmatic skepticism. The "skeptical investigator" was Tony Youens, a British media skeptic with no scientific credentials; his only qualification was that he was a self-proclaimed skeptic. The National Geographic Channel chose to put the full weight of its authority behind Youens's misrepresentation of my research, and I was given no opportunity to reply.

The segment of this show on animal telepathy started with me saying that I had tested the African Gray parrot N'kisi and that he appeared to have telepathic powers (as summarized in Chapter 7). N'kisi's owner, Aimée Morgana, turned down a request to appear because she did not trust National Geographic's motives. So the National Geographic team did a "counter experiment" with an African Gray called Spaulding. One problem with this test was that Spaulding did not show the same kind of telepathic behavior as N'kisi in the first place. In addition, she was tested under stressful conditions that included being moved from her usual place to another part of the house, with strangers from the TV crew all around her. Predictably, the results were no better than chance.

In order to discredit the research with N'kisi, the narrator and Tony Youens then made misleading claims about the statistical analysis of results in the paper that Aimée Morgana and I published in the *Journal of Scientific Exploration*.[34] Based on a long series of controlled tests, this paper provided evidence that a parrot was able to respond telepathically to his owner even though she was in a different room on a different floor and he could not see or hear her.

Here is a transcript of the first of Youens's claims:

**Narrator:** One could argue that perhaps Spaulding's poor performance means that she isn't really telepathic, compared to N'kisi, the bird Rupert Sheldrake tested. But Tony found holes in Sheldrake's experiment too.

**Youens:** The thing that bothers me about the Sheldrake experiment is that if the bird didn't answer, give any credible answer, then they just scrubbed that.

**Narrator:** Sheldrake threw those trials out completely.

*A graphic shows a phrase leaping out of the paper: "They were irrelevant to the analysis."*

The question as to whether trials in which the parrot said nothing should be included in the analysis is a technical one. If the parrot gave no response, it could not be right or wrong, which is why the trial was irrelevant. Omitting trials in which there is no response is standard practice in mainstream research with young children, autistic people, and animals, owing to their limited attention spans. However, one of the referees of our paper explicitly addressed the question of this omission. Here is what he wrote about it, as published in the *Journal of Scientific Exploration* immediately after our paper: "When I originally refereed this article, I was concerned mainly by the omission of the instances in which N'kisi said nothing. It seemed to me that opportunities for him to have had a match, but where he failed, should be counted as failures, regardless of when he said anything or not. I therefore requested data on the omitted cards/phrases, which the authors immediately supplied. I did a permutation test on the entire dataset, and found a *p*-value [probability value] that differed only trivially from the one stated in the article. Although the authors have done an analysis I would not have done (by omitting data), it makes no difference to the results, and so I was happy."

When Youens and the National Geographic producer read our article, they must have known that the omission of the trials in which N'kisi did not respond made no difference. With or without this omission, the results were highly significant statistically. Therefore for them to imply that omitting these trials from the statistical analysis invalidated our results was deliberately misleading.

The program then went on to make a further misleading claim, as follows:

**Narrator:** And if N'kisi didn't come up with words she rarely speaks, he threw out those trials as well.

*A graphic shows another phrase leaping out of the paper saying: "Exclude the eighteen trials involving these images."*

**Youens:** That ultimately stacks the deck. There's no reason he shouldn't still get them, rare words or not, and you've got to include those misses as well as the hits.

*A graphic shows our paper shrinking and spiraling down into a black hole, then disappearing into oblivion with a sucking sound.*

Here is the passage from our paper from which the seemingly incriminating phrase leaped out:

> The list of N'kisi's vocabulary from which the key words had been chosen was not edited for frequency or reliability of use, and it included some words that N'kisi had used only rarely and did not utter at all during this series of trials. These words were "cards," "CD," "computer," "fire," "keys," "teeth," and "TV." There were eighteen trials involving pictures corresponding to these words in which N'kisi could not have scored either a hit or a miss, since he never said these words. In established practices for testing language-using animals, the words tested are typically screened in some way for reliability of production.[35] Perhaps a better way of analyzing the results would be to *exclude the eighteen trials involving these images.* The results of this analysis are shown in Table 4, II. This method reduced the number of misses, and consequently the proportion of N'kisi's hits increased. For example, by the majority scoring method (B), 23 words out of 82 were hits (28 percent). Nevertheless, this method made little difference to the statistical significance of the results, as shown by a comparison of parts I and II of Table 4.[36]

Table 4, I shows the results including *all* key words, and part II shows what happens when the eighteen trials are omitted. There is practically no difference. For example, comparing IC with IIC, the *p*-values arrived at by a randomized permutation analysis are 0.002 and 0.003 respectively, both values being highly significant.

Thus for the narrator and Youens to claim that our analysis of the data was invalid because we omitted rarely said words is deceitful. We were contrasting this method with an all-inclusive method, which we carried out first. Our main conclusions, quoted in the program, were from the all-inclusive method. In other words we showed that the omission of rare words made practically no difference. Our results were highly significant statistically whatever method of analysis we used. The points Youens raised were all fully addressed in our paper. National Geographic knew this, and they deliberately misled viewers in a way that gave a damaging and false impression of our work.

I was given no chance to respond to Youens's comments, but National Geographic's lawyers still claimed that the program "presents the views of each of the parties fairly and in a balanced, professional manner."[37] I alerted other researchers on unexplained phenomena to National Geographic's concept of fairness and advised them to treat any approaches from National Geographic Channel with extreme caution.

A review of the entire *Is It Real?* series by Ted Dace, an independent commentator, helps put this incident in its wider context: "The object of *Is It Real?* is to place its viewers under the purring, hypnotic sway of Science, not science as a method for obtaining reliable knowledge but scientism as a kind of religion that casts out the demons of uncertainty and mystery. Each episode in the series raises the specter of the paranormal only to reveal it as the hallucination of abnormal people. Backed up by a battalion of skeptical commentators . . . *Is It Real?* presents a black and white world of skeptics and believers, and the skeptics turn out to be right every time."[38]

## Richard Dawkins

Richard Dawkins, the author of *The God Delusion*, is a man with a mission—the eradication of religion and superstition, and their total replacement with science and reason. The British TV company Channel 4 repeatedly provided him with a pulpit. In 2006 they broadcast a two-part diatribe against religion called *The Root of All Evil?* followed in 2007 by a sequel called *Enemies of Reason*.

Soon before *Enemies of Reason* was filmed, the production company IWC Media told me that Richard Dawkins wanted to visit me to discuss my research on unexplained abilities of people and animals. They did not tell me that the series was to be called *Enemies of Reason*. I was reluctant to take part because I expected that it would be as one-sided as Dawkins's previous series, and I had already had several negative experiences with TV companies promoting a skeptical agenda, including National Geographic. But the production team's representative, Rebecca Frankel, assured me that they were open-minded, adding, "This documentary, at Channel 4's insistence, will be an entirely more balanced affair than *The Root of All Evil?* was." She told me, "We are very keen for it to be a discussion between two scientists, about scientific modes of inquiry." On the understanding that Dawkins was interested in discussing evidence, and with the written assurance that the material would be edited fairly, I agreed to meet him and we fixed a date.

I was still not sure what to expect. Was he going to be dogmatic, with a mental firewall that blocked out any evidence that went against his beliefs? Or would he be fun to talk to?

Dawkins duly came to call. The Director, Russell Barnes, asked us to stand facing each other; we were filmed with a handheld camera. Dawkins began by saying that he thought we probably agreed about many things, "But what worries me about you is that you are prepared to believe almost anything. Science should be based on the minimum number of beliefs."

I agreed that we had a lot in common, "But what worries me about you is that you come across as dogmatic, giving people a bad impression of science, and putting them off."

Dawkins then said that in a romantic spirit he himself would like to believe in telepathy, but there just wasn't any evidence for it. He dismissed all research on the subject out of hand, without going into any details. He compared the lack of acceptance of telepathy by scientists such as himself with the way in which the echolocation system had been discovered in bats, followed by its rapid acceptance within the scientific community in the 1940s. In fact, as I later discovered, Lazzaro Spallanzani had shown in 1793 that bats rely on hearing to find their way around, but skeptical opponents dismissed his experiments as flawed and helped set back research for more than a century. However, Dawkins recognized that telepathy posed a more radical challenge than echolocation. He said that if it really occurred it would "turn the laws of physics upside down," and added, "Extraordinary claims require extraordinary evidence."

"This depends on what you regard as extraordinary," I replied. "The majority of the population say they have experienced telepathy, especially in connection with telephone calls. In that sense, telepathy is ordinary. The claim that most people are deluded about their own experience is extraordinary. Where is the extraordinary evidence for that?"

He could not produce any evidence at all, apart from generic arguments about the fallibility of human judgment. He also took it for granted that people want to believe in the paranormal because of wishful thinking.

We then agreed that controlled experiments were necessary. I said that this is why I had actually been doing such experiments, including tests to find out if people really could tell who was calling them on the telephone when the caller was selected at random. The results were far above the chance level. The previous week I had sent Dawkins copies of some of my papers in scientific journals so that he could examine some of the data before we met. At this stage he looked uneasy and said, "I don't want to discuss evidence." "Why not?" I asked. He replied, "There isn't time. It's too complicated. And that's not what this program is about." The camera stopped.

Russell Barnes confirmed that he too was not interested in evidence. The film he was making was another Dawkins polemic against irrational beliefs. I said to him, "If you're treating telepathy as an irrational belief, surely evidence about whether it exists or not is essential for the discussion. If telepathy occurs it's not irrational to believe in it. I thought that's what we were going to talk about. I made it clear from the outset that I wasn't interested in taking part in another low-grade debunking exercise."

Dawkins said, "It's not a low-grade debunking exercise; it's a high-grade debunking exercise." I replied that in that case there had been a serious misunderstanding, because I had been assured that this was to be a balanced scientific discussion about evidence. Russell Barnes asked to see the e-mails I had received from his assistant. He read them with obvious dismay and said the assurances she had given me were wrong. The team packed up and left.

Richard Dawkins has long proclaimed his conviction that "The paranormal is bunk. Those who try to sell it to us are fakes and charlatans." *Enemies of Reason* was intended to popularize this belief. But does his crusade really promote the public understanding of science, of which he was the professor at Oxford? Should science be a vehicle of dogma and prejudice, a kind of fundamentalist belief system? Or should it be based on open-minded inquiry into the unknown?

## Skeptical revivalism

The skeptic movement is closely allied to evangelical atheism, and since the turn of the millennium both have undergone a resurgence. One of the most influential figures in this social movement is James Randi, who is greatly admired by Richard Dawkins and other crusading atheists. For the 2009 relaunch of the British *Skeptic* magazine, published by CSICOP/CSI, the cover story was on Randi, and the editor, Chris French, introduced his interview with Randi by writing, "If skeptics were allowed to have patron saints, James Randi would undoubtedly fill that role."[39]

Skeptics admire Randi's belligerent style and his tireless activism in the skeptical cause. Since 2003 he has held an annual gathering of skeptics and atheists in Las Vegas called "The Amazing Meeting," which is like a revivalist rally. Inspirational speakers have included Richard Dawkins, Richard Wiseman, and Michael Shermer. Participants are not just motivated but taught the tricks of the trade. For example, in the 2005 meeting Randi and Shermer gave a seminar entitled "Communicating Skepticism to the Public: A Seminar on Promoting a Scientific View of the World." Attendees were handed a manual that told them how to be a media skeptic: "Becoming an expert is a pretty simple procedure; tell people you're an expert. After you do that, all you have to do is maintain appearances and not give them a reason to believe you're not."

In real science becoming an expert requires qualifications and hard work, but as Randi and Shermer pointed out the rules are different for skeptics. All you need is to form a club with like-minded people: "As head of your local skeptic club, you're entitled to call yourself an authority. If your other two members agree to it, you can be the spokesperson too."[40]

Neither Randi nor Shermer are scientists, and their "scientific view of the world" is a fundamentalist belief system rather than science itself. For decades skeptics have gotten away with deceit, dishonesty, and ignorance by laying claim to the authority of science.[41] Those who disagree with them can then be labeled as ignorant and irrational. But if skeptics want to be taken seriously, then they should be subject to the same kinds of quality control as genuine scientists. In the long term, the cause of science and reason will not be advanced by their unscientific and unreasonable behavior.

## Skeptical credulity

Although committed skeptics see themselves as devoted to science, reason, and critical thinking, they are credulous when it comes to the claims of other skeptics. Many science correspondents share this credulity, which is why the scientific media tend to endow dogmatic skepticism with an authority it does not deserve. For example, when prominent materialists like Lewis Wolpert assert that there is no evidence for telepathy, they are quoted uncritically in newspapers and television shows, as if they know what they are talking about. In fact, they are willfully ignorant of the evidence and are merely expressing their prejudices. They abuse their scientific authority.

The effect of skeptical credulity on science is profound. The great majority of universities neither teach about psychic research or parapsychology nor support research in this field. Since students and professional scientists are not informed about research on these subjects, they are ill-equipped to evaluate the claims of skeptics, and they often take what little information they have from skeptic websites or skeptical propaganda in the media. Serious journalists generally share the prejudices of dogmatic skeptics, or at least defer to them in public for fear of being attacked as ignorant and unscientific. The same is true of most politicians. The result is that there is no public funding for research in these controversial areas.

Meanwhile the popular interest in psychic phenomena is encouraged by downmarket media, reinforcing the belief of skeptics that people who take these phenomena seriously are stupid, ignorant, or deluded.

Skeptical organizations play a useful role in exposing fraudulent psychics and charlatans. But insofar as they inhibit scientific research and inquiry into the unknown, they set back the cause of science and reason rather than promoting it. The present system of science funding reinforces the status quo.

## Open-minded science

Until the early twentieth century, some of the most innovative scientists were amateurs; they did science because it interested them, not because it was a career. Charles Darwin was a striking example. Science is now almost completely institutionalized and professionalized. Career scientists generally lack independence; few can follow their curiosity where it leads. They depend on government, institutional or corporate funding. Their grant applications are peer-reviewed anonymously and committees make the decisions, with the result that caution predominates, and unconventional proposals are passed over in favour of safer, more predictable ones.

Taxpayers fund most of the scientific research carried out in universities and research institutes, but they have almost no say in what gets done. Committees of influential scientists, politicians, and corporate executives determine the priorities. In biology, for example, billions of dollars are spent on sequencing genomes, with results of interest only to a handful of specialists. Meanwhile, there is little or no funding for investigating the topics discussed in this book, such as the ability of animals to give warnings of earthquakes and tsunamis, despite that fact that this research would interest millions of people and might be very useful.

My own proposal for a moderate reform of science funding is that 1 percent of the science budget would be allocated to areas of research proposed by nonprofessionals.[42] The other 99 percent of the funds would be spent as usual. Organizations such as charities, schools, societies, small businesses, and environmental groups would be invited to suggest questions they would like to see answered by research. Within each organization, the very possibility of having a say would probably trigger far-ranging discussions and lead to a sense of involvement. For the first time, an element of democracy would play a part in science. This system could be treated as an experiment and tried out for, say, five years. If it had no useful effects it could be discontinued. If it led to productive research, greater public trust in science, and increased interest among students, the percentage allocated to this fund could be increased.

There are few fields of science today where people outside institutional science can do exciting, hands-on research. But professional scientists have neglected most of the subjects covered in this book, and as a result this field of study is extraordinarily underdeveloped, like the study of magnetism in the seventeenth century, fossils in the eighteenth century, and genetics at the time of Mendel. Precisely because this is a field of inquiry in its infancy, there are remarkable opportunities for original investigations on very low budgets.

## How to take part in research

My colleagues and I would be grateful for reports from readers about their animals and their own experiences. Here are some of the ways that you could help:

1. If you have noticed any behaviour by your animals that you think would contribute to the ongoing research program outlined in this book, please tell us about it by e-mail. We are particularly interested in:
   - Animals finding their owners far away from home
   - Animals that respond to calls from particular people before the phone has been answered
   - Waking sleeping animals by staring at them
   - Warnings by animals of impending epileptic fits or other medical emergencies
   - Warnings of impending disasters
2. Please tell us about your own experiences that suggest telepathic or other invisible interconnections. We are especially interested to hear about:

- Nursing mothers whose milk lets down when their baby is in distress, even if they are miles apart
- An ability to find other people by "feeling" where they are
- A well-developed sense of direction
- Premonitions of earthquakes and other disasters

3. Keep a log of your animal's behavior if it seems to know when a member of the household is coming home. The simplest way to do this is to use a special notebook for this purpose, and note down the time at which the animal reacts, when the person arrives, the time at which he set off, how he traveled, whether or not he was coming at a routine time, and whether or not the people at home knew when to expect him. If the animal fails to respond, this should also be recorded.
4. If your animal responds to calls from people it knows before the phone has been answered, please keep a log of this behaviour, recording the time at which it happens, who it happens with, whether it happens every time that person rings or not, and any other relevant details.
5. Keep a log of your own experiences of picking up people's intentions to call you or send you a message. For example if you feel you know who is calling before you answer the phone or look at the caller ID display, write down your intuition, note what time the call came, and record whether you were right or wrong.
6. Carry out tests with your animals. Throughout this book I have given examples of tests designed to find out whether animals' perceptive behaviour can be explained in terms of habit, routine, and normal sensory information, or whether some other form of communication is involved. More experiments with dogs, cats, parrots, horses, and other species would be very desirable. This research could make an excellent student project. For example, if you have an animal that knows when a member of the family is coming home, arrange for the place it normally waits to be filmed continuously while that person is away from the house, with the time code recorded on the film. Then the person should come home at unusual times, randomly selected, and travel by unfamiliar means, to avoid familiar car sounds. You can adopt a similar procedure if your animal responds to tele-

phone calls. The telephone should be filmed continuously. The person the animal responds to should call at randomly selected times, and other people it does not know should call at other times to find out whether the animal does in fact react selectively.

Please email me your observations, results, or queries through my website www.sheldrake.org.

# Notes

## *Preface*

1. Sheldrake, 1994.

## *Introduction*

1. Serpell, 1986.

2. For a discussion of the mechanistic theory of life and alternatives to it, see Sheldrake, 1990.

3. Pfungst, 1911, p. 10.

## *Chapter 1. The Domestication of Animals*

1. Godwin, 1975; Marx et al., 1988.

2. Leakey and Lewin, 1992; Mithen, 1996.

3. Ehrenreich, 1997.

4. Ibid.

5. Eliade, 1964; Burkert, 1996.

6. Eliade, 1964, p. 94.

7. Masson, 1997.

8. Driscoll and Macdonald, 2010.

9. Morell, 1997.

10. Paxton, 1994.

11. Germonpré et al., 2009.

12. Fiennes and Fiennes, 1968.

13. Serpell, 1983.

14. Ibid.

15. Lindberg et al., 2005.

16. Trut et al., 2004.

17. Galton, 1865.

18. Kerby and Macdonald, 1988.

19. Driscoll et al., 2009.

20. Clutton-Brock, 1981, p. 110.

21. Kiley-Worthington, 1987.

22. For an interesting discussion of the evolution of the Lassie stories, see Garber, 1996.

23. Galton, 1865.

24. Fiennes and Fiennes, 1968.

25. Pet Food Manufacturers Association, United Kingdom, www.pfma.org.uk.

26. American Veterinary Medical Association, www.avma.org.

27. Darwin, 1875.

28. Kiley-Worthington, 1987.

29. Kerby and Macdonald, 1988.

30. Sheldrake, 1988.

31. Francis Huxley, 1959, points out that Darwin's most famous book would more appropriately be called *The Origin of Habits.*

32. Sheldrake, 1981, 1988, 2009.

33. For a mathematical model of communication through a morphic field, see Abraham, 1996.

## *Chapter 2. Dogs*

1. Serpell, 1986.

2. Fogle, 1995, p. 41.

3. Shiu, Munro, and Cox, 1997; Munro, Paul, and Cox, 1997.

4. Boone, 1954, Chapter 7.

5. *Country Life,* November 5, 1999.

6. Serpell, 1986, pp. 103–4.

7. Sheldrake and Smart, 1997; Brown and Sheldrake, 1998; Sheldrake, Lawlor, and Turney, 1998.

8. Readers interested in these surveys can read more about them on my website: www.sheldrake.org.

9. Sheldrake and Smart, 1997; Brown and Sheldrake, 1998; Sheldrake, Lawlor, and Turney, 1998.

10. Matthews, 1994.

11. For the linear correlation between journey time and Jaytee's reaction time, $p < 0.0001$, see Sheldrake and Smart, 1998.

12. On twenty out of fifty-five occasions, Jaytee reacted at the time Pam set out for home, or within two minutes of this time. But sometimes Jaytee reacted before Pam set off and sometimes after: in nine cases he reacted more than three minutes early, and in twenty-six instances he reacted more than three minutes late. Is this variation merely a matter of chance? Or could some of it be due to biases in the way the data were recorded? There could have been at least two sources of bias, working in opposite directions. Some of the data on Jaytee's behavior may be biased toward lateness. If Mr. and Mrs. Smart were not in their sitting room or if they were distracted—for example, by visitors, telephone calls, or television programs—they would not have noticed Jaytee's reactions immediately. Thus, on some of the occasions when Jaytee's reported reactions began after Pam set off to come home, he may in fact have reacted earlier, closer to the time she set off. Conversely, on some of the occasions when Jaytee reacted early, this earliness could be an artifact arising from the way in which Pam's time of setting out was defined. Pam recorded the times at which she actually began her car journey. But sometimes she started getting ready to go ten minutes or more beforehand, taking time to say good-bye to people or chatting as she was leaving. Also, she sometimes thought about leaving before she made a move to do so. If Jaytee was reacting to her intentions, he would have tended to react *before* she set out in the car.

13. Sheldrake and Smart, 1998.

14. Ibid.

15. Ibid.

16. Sheldrake, 1994.

17. Sheldrake and Smart, 1999.

18. Wiseman, Smith, and Milton, 1998.

19. There is no dispute about the facts, but there is a dispute about the interpretation of those facts. Richard Wiseman and Matthew Smith invented a criterion of their own by which to judge Jaytee's success. They decided that Jaytee's "signal" for Pam's return was to be the first time he visited the window for more than two minutes for no obvious external reason. They disregarded all the data subsequent to these so-called signals. In fact, in their experiments

at Pam's parents' flat, although Jaytee went to the window several times during Pam's absence, he spent a far higher proportion of the time at the window when Pam was actually on her way home. On average, Jaytee was at the window only 4 percent of the time for the main part of Pam's absence; in the ten minutes prior to her return, Jaytee was there 48 percent of the time, and while she was actually returning, he was at the window 78 percent of the time. This pattern of results is statistically significant. Wiseman and Smith, however, decided to ignore most of their own data and to disqualify Jaytee if he did not meet their arbitrary two-minute criterion, and thus they were able to claim that Jaytee had failed the test. They announced this conclusion through press releases, television, and newspapers. For a more detailed account, see the Appendix.

20. For more on Wiseman, see www.skeptiko.com.

21. Sheldrake and Smart, 2000.

## *Chapter 3. Cats*

1. Driscoll et al., 2009.
2. Deag, Manning, and Lawrence, 1988.
3. Kerby and Macdonald, 1988.
4. Turner, 1995.

## *Chapter 4. Parrots, Horses, and Other Animals*

1. A statistical analysis using the paired-sample *t* test showed a significance of $p = 0.03$.
2. Barber, 1993.
3. American Veterinary Medical Association, 2010.
4. Story per Anne McLay, historian of the Australian Sisters of Mercy, in a conversation with the author.
5. Van der Post, 1958.
6. Inglis, 1977, p. 18.
7. Lang, 1911.
8. Hygen, 1987.
9. For example, Haynes, 1976, pp. 208–9.
10. Knowles, 1996.

## *Chapter 5. Animals That Comfort and Heal*

1. Partridge, E., 1958, p. 475.

2. For the most influential statement of this point of view, see Dawkins, 1976.

3. The most systematic exposition of this theory is that of Wilson, 1980.

4. For a discussion of the extent to which giving alarm signals can be dangerous for the individual though beneficial to the group, see Ridley, 1996.

5. But if pets and people help each other to survive, then they are genetically codependent and remain so over many generations. They would therefore have been subject to selection for interspecies altruism.

6. Karsh and Turner, 1988.

7. Ibid.

8. Hart, 1995; Dossey, 1997.

9. Lynch and McCarthy, 1969.

10. Friedmann, 1995.

11. Hart, 1995; Rennie, 1997.

12. Hart, 1995.

13. Serpell, 1991.

14. Ibid.

15. Hart, 1995.

16. Dossey, 1997.

17. *NIH News in Health*, February 2009.

18. For example, Paul and Serpell, 1996.

19. For example, Summerfield, 1996.

20. For example, Phear, 1997.

21. Ormerod, 1996.

22. Rennie, 1997.

23. For many examples of healing and comforting by animals, including dogs that visit the sick and dying, see McElroy, 1997.

24. Metzger, 1998.

25. Garber, 1997, pp. 137–38.

26. Edney, 1992.

27. McCormick and McCormick, 1997.

28. Stewart, 1995.

29. Ibid.

30. Masson, 1997.

31. Bossie, 2003.

32. Dosa, 2007.

33. Michell and Rickard, 1982, p. 127.

34. Ibid., p. 128.

## *Chapter 6. Distant Deaths and Accidents*

1. Masson, 1997, p. 144.

2. Bradshaw and Nott, 1995.

3. Morris, 1986, p. 17.

4. Steinhart, 1996, p. 24.

5. Gurney, Myers, and Podmore, 1886; Broad, 1962.

6. Stevenson, 1970.

## *Chapter 7. Picking Up Intentions*

1. Pepperberg, 1999.

2. "Alex, the talking research African gray parrot, passes away," CNN, September 17, 2007. See www.youtube.com/watch?v=c4gTR4+kvcMSeptember.

3. Sheldrake and Morgana, 2003.

## *Chapter 8. Telepathic Calls and Commands*

1. Woodhouse, 1992, p. 54.

2. Sheldrake and Smart, 1997; Sheldrake, Lawlor, and Turney, 1998; Brown and Sheldrake, 1998; Sheldrake, 1998a.

3. Bechterev 1949, p. 175.

4. But some preliminary and rather inconclusive laboratory experiments were carried out with cats by Osis, 1952; and Osis and Foster, 1953.

5. Kiley-Worthington, 1987, pp. 88–89.

6. Roberts, 1996.

7. Blake, 1975, p. 131.

8. Ibid., p. 94.

9. Ibid., p. 129.

10. Smith, 1989.

11. Myers, 1997.

12. St. Barbe Baker, 1942, p. 41.

13. Steiger and Steiger, 1992, p. 16.

14. Sheldrake, 2000, 2003a; Brown and Sheldrake, 2001.

15. Sheldrake, 2003a.
16. Sheldrake and Smart 2003a, b.
17. Sheldrake, Godwin, and Rockell, 2004.
18. Sheldrake and Smart, 2003b.
19. Lobach and Bierman, 2004.
20. Schmidt, S., et al., 2009.
21. Sheldrake and Smart, 2005.
22. Sheldrake, Avraamides, and Novak, 2009.

## *Chapter 9. Animal-to-Animal Telepathy*

1. Hölldobler and Wilson, 2009, p. xx.
2. Von Frisch, 1975, p. 150.
3. Marais, 1973.
4. Sheldrake, 1994.
5. Gordon, 1999, p. 151.
6. Ibid., pp. 152–53.
7. Deborah Gordon, e-mail to the author, October 5, 2004.
8. For a discussion of other possible experiments with termites and ants, see Sheldrake, 1994, Chapter 3.
9. Wilson, 1980, pp. 207–8.
10. Partridge, B., 1981.
11. Ibid., pp. 493–94.
12. Mathematical models of fish schooling have to take into account synergistic or cooperative effects over the entire school, which are one way of representing the field of the school. See, for example, Huth and Wissel, 1992; Niwa, 1994.
13. Selous, 1931, p. 9.
14. Ibid., p. 10.
15. For a summary of recent research on flock behavior and the mathematical modeling of animal groups, see Parrish and Hammer, 1997.
16. Carlson, 2000.
17. Toner and Tu, 1998.
18. Couzin et al., 2005.
19. Couzin, 2007.
20. Ballerini et al., 2008a.
21. Ballerini et al., 2008b.
22. Cavagna et al., 2010.
23. Long, 1910, pp. 101–5.

24. Blake, 1975.
25. Ostrander and Schroeder, 1970.
26. Rogo, 1997.
27. Wylder, 1978.
28. Peoc'h, 1997.

## *Chapter 10. Incredible Journeys*

1. Burnford, 1961.
2. Young, 1995.
3. Lemish, 1996, p. 220.
4. Smith, L., 2002.
5. Haldane, 1997.
6. Campbell, 1999.
7. Read et al., 2007.
8. Herrick, 1922.
9. Schmidt, 1932.
10. Schmidt, 1936, pp. 188–89.
11. Ibid., p. 192.
12. Thomas, 1993, p. 7.
13. Ibid., p. 7.
14. *Out of This World*, BBC One, August 6, 1996.
15. For detailed discussions of morphic fields, see Sheldrake, 1988, 2009.
16. McFarland, 1981.
17. Steinhart, 1995, p. 16.
18. Boitani et al., 1995.
19. Kerby and Macdonald, 1955.
20. Liberg and Sandell, 1955.
21. Carthy, 1963; Matthews, 1968.
22. Matthews, G.V.T., 1968.
23. Carthy, 1963.
24. Gould, 1990.
25. Schmidt-Koenig and Ganzhorn, 1991.
26. Walraff, 1990.
27. Schmidt-Koenig, 1979.
28. Sobel, 1996.
29. Keeton, 1981.
30. Schmidt-Koenig, 1979; Wiltschko, Wiltschko, and Jahnel, 1987.
31. Zapka et al., 2009.

32. Moore, 1988; Walcott, 1991.
33. Van der Post, 1962, p. 235.
34. Forster, 1778.
35. For a summary of research findings, see Baker, 1989.

## *Chapter 11. Migrations and Memory*

1. Gunnarsson et al., 2004.
2. Brower, 1996.
3. Berthold, 1991.
4. Keeton, 1981.
5. Able, 1982.
6. Lohmann, 2010.
7. Wiltschko and Wiltschko, 1995, 1999.
8. Skinner and Porter, 1987.
9. Able and Able, 1996.
10. Sobel, 1996.
11. Hasler, Scholz, and Horrall, 1978.
12. Papi and Luschi, 1996.
13. Ibid.
14. Lohmann, 1992.
15. Jouventin and Weimerskirsch, 1990; Weimerskirsch et al., 1993.
16. Papi and Luschi, 1996.
17. Sheldrake, 1988, 2009.
18. Helbig, 1996.
19. Perdeck, 1958.
20. Ibid.
21. Ibid.
22. Lohmann et al., 2001.
23. Ibid.
24. Thorup et al., 2007, p. 181.
25. Baker, 1980.
26. Helbig, 1996.
27. Bearhop et al., 2005.
28. On this hypothesis, when birds of different migratory races are crossed—blackcaps from Eastern Europe and Western Europe, for example—their offspring would tune in to both sets of migratory habits and would probably be confused. In fact, when such hybrid birds are tested at the beginning of the migratory season to see which way they hop in cages, they show a much

wider variation than birds of the parental races. Caged birds of the eastern race tend to hop toward the southeast, the western race toward the southwest, and the hybrids on average in an intermediate direction, southward (Helbig, 1993, 1996). In real life, if the hybrids persisted on a southward course, they would not follow either of the traditional migratory paths from Europe to Africa, with short sea crossings over the Strait of Gibraltar or the Bosphorus, and would have to find a new wintering place or perish.

29. Bowen and Avise, 1994.

## *Chapter 12. Animals That Know When They Are Nearing Home*

1. Thomas, 1993, p. 143.

## *Chapter 13. Pets Finding Their People Far Away*

1. The details are given in a nineteenth-century account, published in St. Gall, entitled *Zollikofer und sein Hund*, a copy of which was kindly given to me by Professor C. Zollikofer of the University of Zurich, a descendant of the ambassador.

2. Cooper, 1983, p. 149.
3. Geller, 1998.
4. Rhine, 1951.
5. Rhine and Feather, 1962.
6. Whitlock, 1992.
7. Pratt, 1964.
8. Reprinted in *World Farming Newsletter*, 1983.
9. "Cow's Long March," *Soviet Weekly*, January 24, 1987.
10. Long, 1919, p. 95.
11. Ibid., pp. 97–99.

## *Chapter 14. Premonitions of Fits, Comas, and Sudden Deaths*

1. For an enlightening discussion of fear, see Masson, 1996.
2. Hölldobler and Wilson, 1994.
3. Brown, 1975.
4. Dalziel et al., 2003.
5. Chandrasekeran, 1995.

6. Price, 1998.

7. Edney, 1993.

8. Dalziel et al., 2003.

9. Kirton et al., 2004.

10. Smith, 1997.

11 Support Dogs, 21 Jessop's Riverside, Brightside Lane, Sheffield S9 2RX, England.

12. The Delta Society's National Service Dog Center, 875 124th Avenue NE, Suite 101, Bellevue, WA 98005.

13. Chandrasekeran, 1995.

14. Strong et al., 2002.

15. Lim et al., 1992.

16. Chen, M., Daly, M., Natt, Susie, Candie, and Williams, G., 2000. Non-invasive detection of hypoglycemia using a novel, fully biocompatible and patient-friendly alarm system, *British Medical Journal* 321, 2000.

17. Chen et al., 1565–66.

18. Williams and Pembroke, 1989.

19. Ogilvie, 2006.

20. McCulloch et al., 2006.

## *Chapter 15. Forebodings of Earthquakes and Other Disasters*

1. I am grateful to Anna Rigano for these reports. At my request she traveled to Assisi, Foligno, and other earthquake-affected areas of Umbria within three weeks of the September 26, 1997, earthquake. While aftershocks were still occurring, she interviewed dozens of people—including pet owners and veterinarians—about the animal behavior they witnessed prior to the main Assisi earthquake.

2. Quoted in Tributsch, 1982, p. 13.

3. Ibid.

4. Bardens, 1987.

5. Ibid.

6. Wadatsumi, 1995.

7. Sheldrake, 2003a, Chapter 15.

8. Grant and Halliday, 2010.

9. Geller et al., 1997.

10. Tributsch, 1982, p. 9.

11. Ibid., p. 10.

12. Evernden, 1976.

13. Hui, 1996.

14. Hui and Kerr, 1997.

15. Hui, 1996.

16. Otis and Kautz, 1981.

17. Probabilities of the results being due to chance were as low as $p < 0.00005$.

18. Tributsch, 1982, Chapter 5.

19. Otis and Kautz, 1981.

20. Ikeya et al., 1997.

21. Lighthill, 1996.

22. Time Research Institute, Box 620198, Woodside, CA 94962.

23. Ikeya, Takaki, and Takashimizu, 1996; Ikeya, Matsuda, and Yamanaka, 1998; Ikeya, 2004.

24. Newsom and Scott, 1999.

25. Cooper, 1983.

26. Ibid., 128.

27. Peter, 1994.

28. "Belgrade zoo animals provide early bombing warning," Reuters, May 30, 1999.

29. Parson, 1956.

30. Inglis, 1985, p. 74.

31. Radin, 1997, p. 112.

32. Dunne, 1958.

33. Radin, 1997, 2006.

34. For a fuller discussion, see Sheldrake, 2003.

## *Chapter 16. Animal Powers and the Human Mind*

1. Lang, 1911.

2. In our own surveys, more women than men said they'd had a psychic experience, and more women had experienced seemingly telepathic telephone calls: Sheldrake and Smart, 1987; Sheldrake, Lawlor, and Turney, 1998; Brown and Sheldrake, 1998.

3. Baker, 1989.

4. The most striking exception is the pioneering work of Rhine and Feather, 1962. For a review of research by parapsychologists on this subject, see Morris, 1977.

5. Peoc'h, 1988a, b.

6. Peoc'h, 1988c.

7. Peoc'h, 1997.

8. Jahn and Dunne, 1987; Radin, 1997, 2006.

9. For a discussion of the effects of intention and their relationship to positive thinking and prayer, see Sheldrake and Fox, 1996.

10. Sheldrake, 1994; Cottrell, Winer, and Smith, 1996.

11. Elsworthy, 1898.

12. Dundes, 1981.

13. Sheldrake, 2003a.

14. Corbett, 1986; Sheldrake, 2003.

15. Long, 1919.

16. Cottrell et al., 1996.

17. Sheldrake, 1999.

18. Sheldrake, 1994.

19. For a review, see Sheldrake, 2005.

20. Sheldrake, 2003a.

21. Braud, Shafer, and Andrews, 1993a, b; Schlitz and LaBerge, 1994, 1997, Schlitz and Braud, 1997; Delanoy, 2001. A recent meta-analysis of fifteen closed-circuit television staring studies confirmed that there was an overall statistically significant positive effect; see Schmidt et al., 2004.

22. The morphic field concept might account for precognition if it were developed further to take into account the way that waves and vibrations are spread out in time, with no sharp cutoff between past, present, and future, as discussed by Sheldrake, McKenna, and Abraham, 2005.

23. For a discussion of some of these implications, see Sheldrake, McKenna, and Abraham, 2005.

24. Barrow, 1988, p. 361.

25. Davies and Gribbin, 1991, p. 217. A recent experimental development of the principle of nonlocality is the achievement of quantum teleportation (Bouwmeester et al., 1997).

26. For discussions of morphic fields and their implications, see Sheldrake, 1988, 2009.

27. For a discussion of these unexplained human abilities, see Sheldrake, 2003a.

## *Appendix. Controversies and Inquiries*

1. Kuhn, 1970.

2. Carter, 2007.

3. Ibid.

4. *Skeptical Inquirer*, November/December, 2001.

5. To view the author's technical seminar, see www.youtube.com/watch?v=JnA8GUtXpXY.

6. For more on the BBC interview with Maddox, see www.youtube.com/watch?v=aRjQmZLT8bI.

7. Maddox, 1999.

8. For information on Randi, see www.en.wikipedia.org/wiki/James_Randi.

9. *Fate,* October 1981.

10. For a further discussion of Randi's prize offering, see www.skepticalinvestigations.org/New/Organskeptics/index.html#grandprize.

11. For a transcript of the e-mail exchange between Tsakiris and Randi, see www.skeptiko.com/blog/?p=11.

12. Wiseman et al., 1998.

13. Sheldrake and Smart, 1998.

14. Sheldrake and Smart, 2000a.

15. *Equinox: The Secrets of the Psychics,* Channel 4, August 24, 1997.

16. Wiseman et al., 2000.

17. *The Guardian,* March 2, 2004.

18. Carter, 2010.

19. Blackmore, S., 1999.

20. Sheldrake and Smart, 2000, Figure 2, p. 242–43.

21. Including the first sixty minutes of Pam's absence in the analysis by the paired-sample $t$ test, $t = -5.72$, $p = 0.0001$; excluding the first sixty minutes, $t = -5.99$, $p < 0.0001$.

22. For more on Shermer and his Skeptics Society and *Skeptic,* see www.edge.org/q2010/q10_10.html#shermer.

23. *USA Today,* February 26, 2003.

24. Shermer, 2005.

25. For examples of scientists whom Shermer has misrepresented, see www.skepticalinvestigations.org/New/Mediaskeptics?index.html#MichaelShermer.

26. Shermer, 2010.

27. Discovery Channel, August 31, 2001.

28. Whitfield, 2004.

29. For the telepathy debate with Wolpert, see www.skepticalinvestigations.org/New/Audio/telepathy.html.

30. For more on Hardwick's account of the debate, see www.skepticalinvestigations.org/New/Debates/Euroskep_2005.html.

31. For the full text of Ofcom's adjudication, see www.sheldrake.org/D&C/controversies/Ofcom_full.html.

32. In some of the many repeats of the series, the program is called *Is It Real? Animal Oracles.*

33. Dana Kemp, e-mail to the author, February 11, 2005.

34. Sheldrake and Morgana, 2003.

35. Fouts, 1997.

36. Sheldrake and Morgana, 2003, p. 609.

37. Angelo M. Grima, letter to the author, November 11, 2005.

38. To read Dace's review of *Is It Real?*, see www.skepticalinvestigations.org/New/Skepticsmedia/Dace_Isitreal.html.

39. *Skeptic* 22.1, 2009, p. 24.

40. *Skeptic* 22.2, 2010, p. 38.

41. For a penetrating critique on skeptical claims, see Carter, 2007. See also www.skepticalinvestigations.org.

42. Sheldrake, 2003b, 2004.

# References

Able, K. P. 1982. The effects of overcast skies on the orientation of free-flying nocturnal migrants. In F. Papi and H. G. Wallraff, eds. *Avian Navigation.* Springer, Berlin.

Able, K. P., and M. A. Able. 1996. The flexible migratory orientation system of the Savannah sparrow. *Journal of Experimental Biology* 199, 3–8.

Abraham, R. 1996. *Vibrations: Communication through a morphic field.* Visual Math Institute, Santa Cruz.

Abraham, R., T. McKenna, and R. Sheldrake. 1992. *Trialogues at the Edge of the West.* Bear and Company, Santa Fe.

Anderson, A. M. 1982. The great Japanese IQ increase. *Nature* 297, 180–81.

Ash, E. C. 1927. *Dogs: Their History and Development.* Benn, London.

Baker, R. 1980. *The Mystery of Migration.* MacDonald, London.

———. 1989. *Human Navigation and Magnetoreception.* Manchester University Press, England.

Ballerini, M., N. Cabibbo, R. Candelier, A. Cavagna, E. Cisbani, I. Giardina, V. Lecomte, A. Orlandi, G. Parisi, A. Procaccini, M. Viale, and V. Zdravkovic. 2008a. Interaction ruling animal collective behavior depends on topological rather than metric distance: Evidence from a field study. *Proceedings of the National Academy of Sciences,* 105, 1232–37.

Ballerini, M., N. Cabibbo, R. Candelier, A. Cavagna, E. Cisbani, I. Giardina, V. Lecomte, A. Orlandi, G. Parisi, A. Procaccini, M. Viale, and V. Zdravkovic. 2008b. Empirical investigation of starling flocks: A benchmark study in collective animal behaviour. *Animal Behaviour* 76, 201–15.

Barber, T. X. 1993. *The Human Nature of Birds.* St. Martin's Press, New York.

Bardens, D. 1987. *Psychic Animals: An Investigation of Their Secret Powers.* Hale, London.

Barrow, J. 1988. *The World within the World.* Clarendon Press, Oxford.

Bearhop, S., W. Fiedler, R. W. Furness, S. Votier, S. Waldron, J. Newton, G. J. Bowen, P. Berthold, and K. Farnsworth. 2005. Assortative mating as a mechanism for rapid evolution of a migratory divide. *Science* 310, 502–4.

Bechterev, W. 1949; translation. "Direct influence" of a person upon the behaviour of animals. *Journal of Parapsychology* 13, 166–76.

Bedichek, R. 1947; reprint 1961. *Adventures with a Texas Naturalist.* University of Texas Press, Austin.

Berthold, P. 1991. Spatiotemporal programmes and the genetics of orientation. In P. Berthold, ed. *Orientation in Birds.* Birkhäuser, Basel, Switzerland.

Blackmore, S. August 27, 1999. If the truth is out there, we've not found it yet. *Times Higher Education Supplement,* 18.

Blake, H. 1975. *Talking with Horses: A Study of Communication between Man and Horse.* Souvenir Press, London.

Bloxham, J., and D. Gubbins. 1985. The secular variation of the earth's magnetic field. *Nature* 317, 777–81.

Bohm, D., and R. Sheldrake. 1985. Morphogenetic fields and the implicate order. In R. Sheldrake. *A New Science of Life.* 2d ed. Blond, London.

Boitani, L., F. Francisci, P. Ciucci, and G. Andreoli. 1995. Population biology and ecology of feral dogs in central Italy. In J. Serpell, ed. *The Domestic Dog.* Cambridge University Press, England.

Boone, J. A. 1954. *Kinship with All Life.* Harper and Row, New York.

Bossie, L. November 16, 2003. Dog brings comfort to those facing death and those coping with loss. *Sunday Gazette-Mail.*

Bouwmeester, D., J. W. Pan, K. Mattle, M. Eibl, H. Weinfurter, and A. Zellinger. 1997. Experimental quantum teleportation. *Nature* 390, 575–79.

Bowen, B. W., and J. C. Avise. 1994. Tracking turtles through time. *Natural History* 12, 5–6, 38–39.

Bradshaw, J. W. S., and H. M. R. Nott. 1995. Social and communication behaviour of companion dogs. In J. Serpell, ed. *The Domestic Dog.* Cambridge University Press, England.

Braud, W., D. Shafer, and S. Andrews. 1993a. Reactions to an unseen gaze (remote attention): A review, with new data on autonomic staring detection. *Journal of Parapsychology* 57, 373–90.

Braud, W., D. Shafer, and S. Andrews. 1993b. Further studies of autonomic detection of remote staring: replications, new control procedures, and personality correlates. *Journal of Parapsychology* 57, 391–409.

Broad, C. D. 1962. *Lectures on Psychical Research.* Routledge and Kegan Paul, London.

Brower, L. P. 1996. Monarch butterfly orientation. *Journal of Experimental Biology* 199, 93–103.

Brown, D. J., and R. Sheldrake. 1998. Perceptive pets: A survey in northwest California. *Journal of the Society for Psychical Research* 62, 396–406.

Brown, J. L. 1975. *The Evolution of Behavior.* Norton, New York.

Burkert, W. 1996. *The Creation of the Sacred: Tracks of Biology in Early Religions.* Harvard University Press, Cambridge, Mass.

Burnford, S. 1961. *The Incredible Journey.* Hodder and Stoughton, London.

Campbell, B. H. 1999. Homing of translocated brown bears (*Ursos arctus*) in coastal South-Central Alaska. *Northwestern Naturalist* 80, 22–25.

Carter, C. 2007. *Parapsychology and the Skeptics.* Sterling House, Pittsburgh.

———. 2010. "Heads I lose, tails you win": How Richard Wiseman nullifies positive results, and what to do about it. *Journal of the Society for Psychical Research* 74, 156–167.

Carthy, J. D. 1963. *Animal Navigation.* Unwin, London.

Cavagna, A., A. Cimarelli, I. Giardina, G. Parisi, R. Santagati, F. Stefanini, and M. Viale. June 14, 2010. Scale-free correlations in starling flocks. *Proceedings of the National Academy of Sciences* 107, #11, 865–870.

Chandrasekeran, R. July 31, 1995. Epileptic owners swear by seizure-alerting dogs. *Washington Post.*

Clutton-Brock, J. 1981. *Domesticated Animals from Early Times.* Heinemann, London.

Cooper, J. 1983. *Animals in War.* Imperial War Museum, London.

Corbett, J. 1986. *Jim Corbett's India.* Oxford University Press, England.

Cottrell, J. E., G. A. Winer, and M. C. Smith. 1996. Beliefs of children and adults about feeling stares of unseen others. *Developmental Psychology* 32, 50–61.

Couzin, I. D. 2007. Collective minds. *Nature* 445, 715.

Couzin, I. D., J. Krause, N. R. Franks, and S. A. Levin. 2005. Effective leadership and decision-making in animal groups on the move. *Nature* 433, 513–16.

Dalziel, D. J., B. M. Uthman, S. P. McGorray, and R. L. Reep. 2003. Seizure-alert dogs: A review and preliminary study. *Seizure* 12, 115–20.

Darwin, C. 1875. *The Variation of Animals and Plants under Domestication.* Murray, London.

Davies, P., and J. Gribbin. 1991. *The Matter Myth.* Viking, London.

Dawkins, R. 2006. *The God Delusion.* Boston: Houghton Mifflin Harcourt.

———. 1976. *The Selfish Gene.* Oxford University Press, England.

Deag, J. M., A. Manning, and C. A. Lawrence. 1988. Factors influencing the mother-kitten relationship. In D. C. Turner and P. Bateson, eds. *The Domestic Cat.* Cambridge University Press, England.

Delanoy, D. 2001. Anomalous psychophysiological responses to remote cognition: The DMILS studies. *European Journal of Parapsychology* 16, 30–41.

Dosa, D. 2007. A day in the life of Oscar the cat. *New England Journal of Medicine* 357, 328–29.

Dossey, L. 1997. The healing power of pets: A look at animal-assisted therapy. *Alternative Therapies* 3, 8–15.

Driscoll, C. A., D. W. Macdonald, and S. J. O'Brien. 2009. From wild animals to domestic pets, an evolutionary view of domestication. *Proceedings of the National Academy of Sciences* 106 (Supplement 1), 9971–76.

Driscoll, C. A., and D. W. Macdonald. 2010. Top dogs: Wolf domestication and wealth. *Journal of Biology* 9, 10.

Dundes, A., ed. 1981. *The Evil Eye: A Casebook.* University of Wisconsin Press, Madison.

Dunne, J. W. 1958. *An Experiment with Time.* 3d ed. Faber and Faber, London.

Dürr, H. P. 1997. Sheldrakes Vorstellungen aus dem Blickwinkel der modernen Physik. In H. P. Dürr and F. T. Gottwald, eds. *Rupert Sheldrake in der Diskussion.* Scherz Verlag, Bern, Switzerland.

Edney, A. T. B. 1992. Companion animals and human health. *Veterinary Record* 130, 285–87.

———. 1993. Dogs and human epilepsy. *Veterinary Record* 132, 337–38.

Ehrenreich, B. 1997. *Blood Rites.* Metropolitan Books, New York.

Eliade, M. 1964. *Shamanism: Archaic Techniques of Ecstasy.* Princeton University Press, Princeton, N.J.

Elsworthy, F. 1898. *The Evil Eye.* Murray, London.

Evernden, J. R., ed. September 23–24, 1976. *Abnormal Animal Behavior Prior to Earthquakes.* U.S. National Earthquake Hazards Reduction Program, Conference. Menlo Park, Calif.

Fiennes, R., and A. Fiennes. 1968. *The Natural History of the Dog.* Weidenfeld and Nicholson, London.

Flynn, J. R. 1983. Now the great augmentaton of the American IQ. *Nature* 301, 655.

———. 1984. The mean IQ of Americans: Massive gains 1932 to 1978. *Psychological Bulletin* 95, 29–51.

———. 1987. Massive IQ gains in 14 nations. *Psychological Bulletin* 101, 171–91.

Fogle, B. 1994. Unexpected dog ownership findings from Eastern Europe. *Anthrozoös* 7, 270.

———. 1995. *The Encyclopedia of the Dog.* Dorling Kindersley, London.

Forster, J. R. 1778. *Observations Made during a Voyage around the World.* Robinson, London.

Fouts, R. 1997. *Next of Kin: My Conversations with Chimpanzees.* Avon, New York.

Friedmann, E. 1995. The role of pets in enhancing human well-being: Physiological effects. In I. Robinson, ed. *The Waltham Book of Human-Animal Interaction: Benefits and Responsibilities of Pet Ownership.* Pergamon Press, Oxford, England.

Galton, F. 1865. The first steps towards the domestication of animals. *Transactions of the Ethnological Society of London.* New Series 3, 122–38.

Garber, M. 1996. *Dog Love.* Hamish Hamilton, London.

Geller, R. J., D. D. Jackson, Y. Y. Kagan, and F. Mulargia. 1997. Earthquakes cannot be predicted. *Science* 275, 1616–17.

Geller, U. May 27, 1998. Uri Geller's weird web. *Times* (London).

Germonpré, M., M. V. Sablin, R. E. Stevens, R. E. M. Hedges, M. Hofreiter, M. Stiller, and V. R. Després. 2009. Fossil dogs and wolves from Palaeolithic sites in Belgium, the Ukraine and Russia: Osteometry, ancient DNA and stable isotopes. *Journal of Archaeological Science* 36, 473–90.

Godwin, R. D. 1975. Trends in the ownership of domestic pets in Great Britain. In R. S. Anderson, ed. *Pet Animals and Society.* Balliere Tindall, London.

Gordon, D. 1999. *Ants at Work: How an Insect Society Is Organized.* Simon & Schuster, New York.

Goswami, A. 1997. Eine quantentheoretische Erklärung von Sheldrakes morphischer Resonanz. In H. P. Dürr and F. T. Gottwald, eds. *Rupert Sheldrake in der Diskussion.* Scherz Verlag, Bern, Switzerland.

Gould, J. L. 1990. Why birds (still) fly south. *Nature* 347, 331.

Grant, R., and T. Halliday. March 31, 2010. Predicting the unpredictable: Evidence of pre-seismic anticipatory behaviour in the common toad. *Journal of Zoology.*

Gunnarsson, T. G., J. A. Gill, T. Sigurbjörnsson, and W. J. Sutherland. 2004. Arrival synchrony in migratory birds. *Nature* 431, 646.

Gurney, E., F. Myers, and F. Podmore. 1886. *Phantasms of the Living.* Trubner, London.

Haldane, A. R. B. 1997. *The Drove Roads of Scotland.* Birlinn, Edinburgh.

Hart, L. A. 1995. Dogs as human companions: A review of the relationship. In J. Serpell, ed. *The Domestic Dog.* Cambridge University Press, England.

Hasler, A. D., A. T. Scholz, and R. M. Horrall. 1978. Olfactory imprinting and homing in salmon. *American Scientist* 66, 347–55.

Haynes, R. 1976. *The Seeing Eye, The Seeing I.* Hutchinson, London.

Helbig, A. J. 1993. What do we know about the genetic basis of bird orientation? *Journal of Navigation* 46, 376–82.

———. 1996. Genetic basis, mode of inheritance and evolutionary changes of migratory direction in Palearctic warblers. *Journal of Experimental Biology* 199, 49–55.

Herrick, F. H. 1922. Homing powers of the cat. *Science Monthly* 14, 526–39.

Hölldobler, B., and E. O. Wilson. 1994. *Journey to the Ants: A Story of Scientific Exploration.* Harvard University Press, Cambridge, Mass.

Hölldobler, B., and E. O. Wilson. 2009. *The Superorganism: The Beauty, Elegance, and Strangeness of Insect Societies.* Norton, New York.

Horgan, J. November 1995. Get smart, take a test: A long-term rise in IQ scores baffles intelligence experts. *Scientific American*, 10–11.

Hui, L. 1996. China's campaign to predict quakes. *Science* 273, 1484–86.

Hui, L., and R. H. Kerr. 1997. Warnings precede Chinese tremblors. *Science* 276, 526.

Huth, A., and C. Wissel. 1992. The simulation of the movement of fish schools. *Journal of Theoretical Biology* 156, 365–85.

Huxley, F. 1959. Charles Darwin: Life and habit. *The American Scholar* Fall-Winter, 1–19.

Hygen, G. 1987. *Vardøger: Vårt Paranormale Nasjonalfenomen.* Cappelens Forlag, Oslo, Norway.

Ikeya, M. 2004. *Earthquakes and Animals: From Folk Legends to Science.* World Scientific Publishing Co., Singapore.

Ikeya, M., T. Matsuda, and Y. Yamanaka. 1998. Reproduction of mimosa and clock anomalies before earthquakes. *Proceedings of the Japanese Academy* 74B, 60–64.

Ikeya, M., S. Takaki, H. Matsumoto, A. Tani, and T. Komatsu. 1997. Pulsed charge model of fault behavior producing seismic electrical signals. *Journal of Circuits, Systems and Computers* 7, 153–64.

Ikeya, M., S. Takaki, and T. Takashimizu. 1996. Electric shocks resulting in seismic animal anomalous behavior. *Journal of the Physical Society of Japan* 65, 710–12.

Inglis, B. 1977. *Natural and Supernatural.* Hodder and Stoughton, London.

———. 1985. *The Paranormal: An Encyclopedia of Psychic Phenomena.* Granada, London.

Jahn, R. J., and B. Dunne. 1987. *Margins of Reality.* Harcourt Brace, New York.

Jouventin, P., and H. Weimerskirsch. 1990. Satellite tracking of wandering albatrosses. *Nature* 343, 746–48.

Karsh, E. B., and D. C. Turner. 1988. The human-cat relationship. In D. C. Turner and P. Bateson, eds. *The Domestic Cat.* Cambridge University Press, England.

Keeton, W. T. 1981. Orientation and navigation of birds. In D. J. Aidley, ed. *Animal Migration.* Society for Experimental Biology Seminar Series 13, Cambridge University Press, England.

Keller, O. 1913. *Antike Tierwelt.* Engelmann, Leipzig.

Kerby, G., and D. W. Macdonald. 1988. Cat society and the consequences of colony size. In D. C. Turner and P. Bateson, eds. *The Domestic Cat.* Cambridge University Press, England.

Kiley-Worthington, M. 1987. *The Behaviour of Horses.* J. A. Allen, London.

Kirton, A., E. Wirrell, J. Zhang, and L. Hamiwka. 2004. Seizure-alerting and -response behaviors in dogs living with epileptic children. *Neurology* 62, 2303–5.

Knowles, O. S. 1996. Letter. *Psi Researcher* 21, 24.

Kuhn, T. S. 1970. *The Structure of Scientific Revolutions.* Chicago University Press, Chicago.

Lang, A. 1911. Second sight. *Encyclopaedia Britannica,* 11th ed. Cambridge University Press, England.

Leakey, R., and R. Lewin. 1992. *Origins Reconsidered.* Little Brown, London.

Lemish, G. H. 1996. *War Dogs: Canines in Combat.* Brassey, Washington, D.C.

Liberg, O., and M. Sandell. 1955. Spatial organization and reproductive tactics in the domestic cat and other felids. In D. C. Turner and P. Bateson, eds. *The Domestic Cat.* Cambridge University Press, England.

Lighthill, J., ed. 1996. *A Critical Review of VAN.* World Scientific, Singapore.

Lim, K., A. Wilcox, M. Fisher, and C. J. Burns-Cox. 1992. Type 1 diabetics and their pets. *Diabetic Medicine* 9, Supp. 2, S3.

Lindberg, J., S. Bjornerfeldt, P. Saetre, K. Svartberg, B. Seehuus, M. Bakken, C. Vila, and E. Jazin. 2005. Selection for tameness has changed brain gene expression in silver foxes. *Current Biology* 22, 915–16.

Lobach, E., and D. J. Bierman. 2004. Who's calling at this hour? Local sidereal time and telephone telepathy. *Proceedings of the Parapsychological Association Annual Convention,* 2004, 91–97.

Lohmann, K. J. January 1992. How sea turtles navigate. *Scientific American,* 75–82.

———. 2010. Magnetic-field perception. *Nature* 464, 1140–42.

Lohmann, K. J., S. D. Cain, S. A. Dodge, and C. M. F. Lohmann. 2001. Regional magnetic fields as navigational markers for sea turtles. *Science* 294, 362–68.

Long, W. 1919. *How Animals Talk.* Harper, New York. New edition, 2005, Bear and Co., Rochester, VT.

Lynch, J. J., and J. F. McCarthy. 1969. Social responding in dogs: Heart rate changes to a person. *Psychophysiology* 5, 389–93.

Maddox, J. 1999. Dogs, telepathy and quantum mechanics. *Nature* 401, 849.

Marais, E. 1973. *The Soul of the White Ant.* Penguin, Harmondsworth, England.

Marx, M. B., L. Stallones, T. F. Garrity, and T. P. Johnson. 1988. *Anthrozoös* 2, 33–37.

Masson, J. M. 1996. *When Elephants Weep.* Delta, New York.

———. 1997. *Dogs Never Lie about Love.* Cape, London.

Matthews, G. V. T. 1968. *Bird Navigation,* 2d ed. Cambridge University Press, England.

Matthews, R. April 24, 1994. Animal magic or mysterious sixth sense? *Sunday Telegraph* (London).

———. January 15, 1995. Psychic dog gives scientist a lead. *Sunday Telegraph* (London).

McCormick, A., and D. McCormick. 1997. *Horse Sense and the Human Heart.* Health Communications, Deerfield Beach, Fla.

McElroy, S. C. 1997. *Animals as Teachers and Healers.* Ballantine, New York.

McFarland, D., ed. 1981. Navigation. *Oxford Companion to Animal Behaviour.* Oxford University Press, England.

Metzger, D. 1998. Coming home. In B. Peterson, D. Metzger, and L. Hogan, eds. *Intimate Nature: The Bond between Women and Animals.* Ballantine, New York.

Michell, J., and J. M. Rickard. 1982. *Living Wonders: Mysteries and Curiosities of the Animal World.* Thames and Hudson, London.

Mikulecky, M. 1996. Sheldrake versus Rose. *Biology Forum* 89, 469–78.

Mithen, S. 1996. *The Prehistory of the Mind: A Search for the Origins of Art, Religion and Science.* Thames and Hudson, London.

Moore, B. R. 1988. Magnetic fields and orientation in homing pigeons: The experiments of the late W. T. Keeton. *Proceedings of the National Academy of Sciences* 85, 4907–9.

Morell, V. 1997. The origin of dogs: Running with the wolves. *Science* 276, 1647–48.

Morris, D. 1986. *Dogwatching.* Cape, London.

Morris, R. L. 1977. Parapsychology, biology and ANPSI. In B. B. Wolman, ed. *Handbook of Parapsychology.* Van Nostrand Reinhold, New York.

Munro, K. J., B. Paul, and C. L. Cox. 1997. Normative auditory brainstem response data for bone conduction in the dog. *Journal of Small Animal Practice* 38, 353–56.

Myers, A. 1997. *Communicating with Animals.* Contemporary Books, Chicago.

Niwa, H. S. 1994. Self-organizing dynamic model of fish schooling. *Journal of Theoretical Biology* 171, 123–36.

Ormerod, E. 1996. Pet programmes in prisons. *Society for Companion Animal Studies Journal* 8 (4), 1–3.

Osis, K. 1952. A test of the occurrence of a psi effect between man and the cat. *Journal of Parapsychology* 16, 233–56.

Osis, K., and E. B. Forster. 1953. A test of ESP in cats. *Journal of Parapsychology* 17, 168–86.

Ostrander, S., and L. Schroeder. 1970. *Psychic Discoveries behind the Iron Curtain.* Abacus Books, London.

Otis, L. S., and W. H. Kautz. 1981. *Biological premonitors of earthquakes: A validation study.* Annual report prepared for the U.S. Geological Service.

Papi, F., and P. Luschi. 1996. Pinpointing "Isla Meta": The case of sea turtles and albatrosses. *Journal of Experimental Biology* 199, 65–71.

Parrish, J. K., and W. M. Hammer, eds. 1997. *Animal Groups in Three Dimensions.* Cambridge University Press, England.

Parson, N. A. 1956. *Guided Missiles in War and Peace.* Harvard University Press, Cambridge, Mass.

Partridge, B. 1981. Schooling. In D. McFarland, ed. *The Oxford Companion to Animal Behaviour.* Oxford University Press, England.

Partridge, E. 1958. *Origins: A Short Etymological Dictionary of Modern English.* Routledge and Kegan Paul, London.

Paul, E. S., and J. A. Serpell. 1996. Obtaining a new pet dog: Effects on middle childhood children and their families. *Applied Animal Behaviour Science* 47, 17–29.

Paxton, D. 1994. Urban animal management. *Proceedings of the Third National Conference on Urban Animal Management in Australia.* Australian Veterinary Association, Canberra.

Peoc'h, R. 1988a. Action psychocinetique des poussins sur un générateur aléatoire. *Revue Française de Psychotronique* 1, 11–24.

———. 1988b. Chicken imprinting and the tychoscope: An ANPSI experiment. *Journal of the Society for Psychical Research* 55, 1–9.

———. 1988c. Psychokinetic action of young chicks on an illuminated source. *Journal of Scientific Exploration* 9, 223–29.

———. 1997a. Telepathy experiments between rabbits. *Fondation Odier de Psycho-Physique Bulletin* 3, 25–28.

———. 1997b. Telekinesis experiments with rabbits. *Fondation Odier de Psycho-Physique Bulletin* 3, 28–36.

Pepperberg, I. 1999. *The Alex Studies: Cognitive and Communicative Abilities of Grey Parrots.* Harvard University Press, Cambridge, Mass.

Perdeck, A. C. 1958. Two types of orientation in migrating starlings and chaffinches as revealed by displacement experiments. *Ardea* 46, 1–37.

Peter, M. November 26, 1994. Fliegerangriff! Tauben schlugen Alarm. Kronen Zeitung, Vienna.

Pfungst, O. 1911. *Clever Hans: A Contribution to Experimental Animal and Human Psychology.* Henry Holt, New York.

Phear, D. 1997. A study of animal companionship in a day hospice. *Society for Companion Animal Studies Journal* 9 (1), 1–3.

Pinker, S. 1994. *The Language Instinct.* Penguin, London.

Pratt, J. G. 1964. *Parapsychology: An Insider's View of ESP.* W. H. Allen, London.

Price, P. February 21, 1998. Back from the dead. *Times* (London).

Radin, D. 1997. *The Conscious Universe: The Scientific Truth of Psychic Phenomena.* Harper, San Francisco.

———. 2006. *Entangled Minds: Extrasensory Experiences in a Quantum Reality.* Paraview Pocket Books, New York.

Read, M. A., G. C. Grigg, S. R. Irwin, D. Shanaham, and C. E. Franklin. 2007. Satellite tracking reveals long distance coastal travel and homing by translocated estuarine crocodiles. *Public Library of Science One,* 2(9), e949.

Rennie, A. 1997. The therapeutic relationship between animals and humans. *Society for Companion Animal Studies Journal* 9 (4), 1–4.

Rhine, J. B. 1951. The present outlook on the question of psi in animals. *Journal of Parapsychology* 15, 230–51.

Rhine, J. B., and S. R. Feather. 1962. The study of cases of psi-trailing in animals. *Journal of Parapsychology* 16, 1–22.

Ridley, M. 1996. *The Origins of Virtue.* Viking, London.

Roberts, M. 1996. *The Man Who Listens to Horses.* Hutchinson, London.

Rogo, D. S. 1997. Do animals have ESP? In *Psychic Pets and Spirit Animals.* Llewellyn, St. Paul, Minn.

Schechter, B. January 23, 1999. Birds of a feather. *New Scientist* 30–33.

Schlitz, M. J., and S. LaBerge. 1997. Covert observation increases skin conductance in subjects unaware of when they are being observed: A replication. *Journal of Parapsychology* 61, 185–96.

Schmidt, B. 1932. Vorläufiges Versuchsergebnis über das hundliche Orientierungsproblem. *Zeitschrift für Hunderforschung* 2, 133–56.

———. 1936. *Interviewing Animals.* Allen and Unwin, London.

Schmidt, S., R. Schneider, J. Utts, and H. Walach. 2004. Distant intentionality and the feeling of being stared at: Two meta-analyses. *British Journal of Psychology* 95, 235–47.

Schmidt, S., D. Erath, V. Ivanova, and H. Walach. 2009. Do you know who is calling? Experiments on anomalous cognition in phone call receivers. *The Open Psychology Journal* 2, 12–18.

Schmidt-Koenig, K. 1979. *Avian Orientation and Navigation*. Academic Press, London.

Schmidt-Koenig, K., and J. U. Ganzhorn. 1991. On the problem of bird navigation. In P. P. G. Bateson and P. H. Klopfer, eds. *Perspectives in Ethology*, vol. 9. Plenum Press, New York.

Selous, E. 1931. *Thought Transference or What? in Birds.* Constable, London.

Serpell, J. 1983. Best friend or worst enemy: Cross-cultural variation in attitudes to the domestic dog. *Proceedings of the 1983 International Symposium of the Human-Pet Relationship.* Austrian Academy of Sciences, Vienna.

———. 1986. *In the Company of Animals.* Cambridge University Press, England.

———. 1991. Beneficial effects of pet ownership on some aspects of human health and behaviour. *Journal of the Royal Society of Medicine* 84, 717–20.

Sheldrake, R. 1985. *A New Science of Life: The Hypothesis of Formative Causation*, 2d ed. Blond and Briggs, London.

———. 1988a. *The Presence of the Past: Morphic Resonance and the Habits of Nature.* Times Books, New York.

———. February 11, 1988b. Cattle fooled by phoney grids. *New Scientist*, 65.

———. 1990. *The Rebirth of Nature: The Greening of Science and God.* Bantam, New York.

———. 1994. *Seven Experiments That Could Change the World: A Do-It-Yourself Guide to Revolutionary Science.* Fourth Estate, London.

———. 1998. Perceptive pets with puzzling powers: Three surveys. *International Society for Anthrozoology Newsletter* 15, 2–5.

———. 1999. Commentary on a paper by Wiseman, Smith and Milton on the "psychic pet" phenomenon. *Journal of the Society for Psychical Research* 63, 306–11.

———. 2003a. *The Sense of Being Stared At, and Other Aspects of the Extended Mind.* Crown, New York.

———. April 19, 2003b. Set them free. *New Scientist*, 23.

———. 2004. Public participation: Let the people pick projects. *Nature* 432, 271.

———. 2005. The sense of being stared at. Part 1. Is it real or illusory? *Journal of Consciousness Studies* 12, 10–31.

———. 2009. *Morphic Resonance: The Nature of Formative Causation* (A new edition of *A New Science of Life*). Park Street Press, Rochester, Vt.

Sheldrake, R., L. Avraamides, and M. Novak. 2009. Sensing the sending of SMS messages: An automated test. *Explore: The Journal of Science and Healing* 5, 272–76.

Sheldrake, R., and M. Fox. 1996. *Natural Grace: Dialogues on Science and Spirituality.* Bloomsbury, London.

Sheldrake, R., H. Godwin, and S. Rockell. 2004. A filmed experiment on telephone telepathy with the Nolan sisters. *Journal of the Society for Psychical Research* 68, 168–72.

Sheldrake, R., C. Lawlor, and J. Turney. 1998. Perceptive pets: A survey in London. *Biology Forum* 91, 57–74.

Sheldrake, R., T. McKenna, and R. Abraham. 2005. *The Evolutionary Mind.* Monkfish Books, NY.

Sheldrake, R., and A. Morgana. 2003. Testing a language-using parrot for telepathy. *Journal of Scientific Exploration* 17, 601–16.

Sheldrake, R., and P. Smart. 1997. Psychic pets: A survey in northwest England. *Journal of the Society for Psychical Research* 61, 353–64.

Sheldrake, R., and P. Smart. 1998. A dog that seems to know when its owner is returning: Preliminary investigations. *Journal of the Society for Psychical Research* 62, 220–32.

Sheldrake, R., and P. Smart. 2000a. A dog that seems to know when its owner is returning: Videotaped experiments. *Journal of Scientific Exploration* 14, 233–55.

Sheldrake, R., and P. Smart. 2000b. Testing a return-anticipating dog, Kane. *Anthrozoös* 13, 203–11.

Sheldrake, R., and P. Smart. 2003a. Experimental tests for telephone telepathy. *Journal of the Society for Psychical Research* 67, 184–99.

Sheldrake, R., and P. Smart. 2003b. Videotaped experiments on telephone telepathy. *Journal of Parapsychology* 67, 187–206.

Sheldrake, R., and P. Smart. 2005. Testing for telepathy in connection with emails. *Perceptual and Motor Skills* 101, 771–86.

Shermer, M. November 2005. Rupert's resonance. *Scientific American.*

———. May 15, 2010. I am a skeptic, but I'm not a denier. *New Scientist* 36–37.

Shiu, J. N., K. J. Munro, and C. L. Cox. 1997. Normative auditory brainstem response data for hearing threshold and neuro-otiological diagnosis in the dog. *Journal of Small Animal Practice* 38, 103–7.

Skinner, B. J., and S. C. Porter. 1987. *Physical Geology.* John Wiley, New York.

Smith, H. June 8, 1997. My psychic bunny's a lifesaver. *News of the World Magazine.*

Smith, L. July 24, 2002. Dog overboard swims home across the Solent. *Times* (London).

Smith, P. 1989. *Animal Talk: Interspecies Telepathic Communication.* Pegasus, Point Reyes, Calif.

Sobel, D. 1996. *Longitude.* Fourth Estate, London.

St. Barbe Baker, R. 1942. *African Drums.* Lindsay Drummond, London.

Steiger, B., and S. H. Steiger. 1992. *Strange Powers of Pets.* Donald Fine, New York.

Steinhart, P. 1995. *The Company of Wolves.* Knopf, New York.

Stevenson, I. 1970. *Telepathic Impressions.* University Press of Virginia, Charlottesville.

Stewart, M. 1995. Dogs as counselors? *The Society for Companion Animal Studies Journal* 7 (4), 1–4.

Strong, V., S. Brown, M. Huyton, and H. Coyle. 2002. Effect of trained seizure alert dogs on frequency of tonic-clonic seizures. *Seizure* 11, 402–5.

Summerfield, H. 1996. Pets as therapy. *Society for Companion Animal Studies Journal* 8 (4), 9.

Thom, R. 1975. *Structural Stability and Morphogenesis.* Benjamin, Reading, Mass.

———. 1983. *Mathematical Models of Morphogenesis.* Horwood, Chichester, England.

Thomas, E. M. 1993. *The Hidden Life of Dogs.* Houghton Mifflin, Boston.

Toner, J., and Y. Tu. 1998. Flocks, herds, and schools: A quantitative theory of flocking. *Physical Review E* 558, 4828–58.

Tributsch, H. 1982. *When the Snakes Awake.* MIT Press, Cambridge, Mass.

Trut, I. N., I. Z. Plyusnina, and I. N. Oskina. 2004. An experiment on fox domestication and debatable issues on the evolution of the dog. *Genetika* 40, 794–807.

Turner, D. C. 1995. The human-cat relationship. In I. Robinson, ed. *The Waltham Book of Human-Animal Interaction.* Pergamon, Oxford, England.

Van der Post, L. 1962. *The Lost World of the Kalahari.* Penguin, London.

Von Frisch, K. 1975. *Animal Architecture,* Hutchinson, London.

Wadatsumi, K. 1995. *Witnesses 1519 Prior to Earthquake.* Tokyo Publishers, Tokyo (in Japanese).

Waddington, C. H. 1957. *The Strategy of the Genes.* Allen and Unwin, London.

Walcott, C. 1991. Magnetic maps in pigeons. In P. Berthold, ed., *Orientation in Birds.* Birkhäuser, Basel, Switzerland.

Walraff, H. G. 1990. Navigation by homing pigeons. *Ethology, Ecology and Evolution* 2, 81–115.

Weimerskirsch, H., M. Salamolard, F. Sarrazin, and P. Jouventin. 1993. Foraging strategy of wandering albatrosses through the breeding season: A study using satellite telemetry. *Auk* 110, 325–41.

Whitfield, J. 2004. Telepathy charm seduces audience at paranormal debate. *Nature* 427, 277.

Whitlock, R. December 4, 1992. How do they do it? *Guardian Weekly* (London).

Williams, H., and A. Pembroke. April 1989. Sniffer dogs in the melanoma clinic? *The Lancet,* 734.

Wilson, E. O. 1971. *The Social Insects.* Harvard University Press, Cambridge, Mass.

———. 1980. *Sociobiology.* Harvard University Press, Cambridge, Mass.

Wiltschko, R., and W. Wiltschko. 1995a. Das Orientierungssystem der Vögel: I. Kompassmechanismen. *Journal für Ornithologie* 140, 1–40.

Wiltschko, R., and W. Wiltschko. 1995b. *Magnetic Orientation in Animals.* Springer Verlag, Berlin.

Wiltschko, W., R. Wiltschko, and M. Jahnel. 1987. The orientation behaviour of anosmic pigeons in Frankfurt, Germany. *Animal Behaviour* 35, 1328–33.

Wiseman, R., and M. Schlitz. 1997. Experimenter effects and the remote detection of staring. *Journal of Parapsychology* 61, 197–207.

Wiseman, R., M. Smith, and J. Milton. 1998. Can animals detect when their owners are returning home? An experimental test of the 'psychic pet' phenomenon. *British Journal of Psychology* 89, 453–62.

Wiseman, R., M. Smith, and J. Milton. 2000. The 'psychic pet' phenomenon: A reply to Rupert Sheldrake. *Journal of the Society for Psychical Research* 64, 46–49.

Woodhouse, B. 1992. *How Your Dog Thinks.* Ringpress, Letchworth, England.

Wylder, J. 1978. *Psychic Pets: The Secret World of Animals.* Stonehill, New York.

Young, R. September 9, 1995. Dog walks 60 miles home to its master. *Times* (London).

Zapka, M., D. Heyers, C. M. Hein, S. Engels, N. Schneider, J. Hans, S. Weiler, D. Dreyer, D. Kishkines, J. M. Wild, and H. Mouritsen. 2009. Visual but not trigeminal mediation of magnetic compass information in a migratory bird. *Nature* 461, 1274–77.

# Index

Most people cited in this book were happy for their real names to be used; but some asked to be referred to by pseudonyms, indicated by asterisks.
*Page numbers of illustrations appear in italics.*

## *About the Author*

Rupert Sheldrake was born in Newark-on-Trent, England. He studied natural sciences at Cambridge and philosophy at Harvard, then took a Ph.D. at Cambridge and became Director of Studies and Fellow at Clare College, Cambridge. He is the author of more than eighty technical papers in scientific journals and has published ten books. From 2005 to 2010 he was Director of the Perrott-Warrick Project for research on the unexplained abilities of people and animals, funded by Trinity College, Cambridge. He is a Fellow of the Institute of Noetic Sciences in California, and a visiting Professor at the Graduate Institute in Connecticut. He is married, has two sons, and lives in London. His website is www.sheldrake.org.